India Studies in Business and Economics

The Indian economy is considered to be one of the fastest growing economies of the world with India amongst the most important G-20 economies. Ever since the Indian economy made its presence felt on the global platform, the research community is now even more interested in studying and analyzing what India has to offer. This series aims to bring forth the latest studies and research about India from the areas of economics, business, and management science. The titles featured in this series will present rigorous empirical research, often accompanied by policy recommendations, evoke and evaluate various aspects of the economy and the business and management landscape in India, with a special focus on India's relationship with the world in terms of business and trade.

More information about this series at http://www.springer.com/series/11234

Abhijit Das · Joyashree Roy
Sayantan Chakrabarti

Socio-Economic Analysis of Arsenic Contamination of Groundwater in West Bengal

Springer

Abhijit Das
Vijaygarh Jyotish Ray College
Kolkata
India

Sayantan Chakrabarti
Kolkata
India

Joyashree Roy
Department of Economics
Jadavpur University
Kolkata
India

ISSN 2198-0012 ISSN 2198-0020 (electronic)
India Studies in Business and Economics
ISBN 978-981-10-9227-5 ISBN 978-981-10-0682-1 (eBook)
DOI 10.1007/978-981-10-0682-1

Printed on acid-free paper

This Springer imprint is published by SpringerNature
The registered company is Springer Science+Business Media Singapore Pte Ltd.

Preface

We decided to get this book published to help the larger audience make sense of the prevalent diverse, and at times conflicting and confusing, paradigms on arsenic contaminated groundwater use in India. The arsenic contamination of groundwater in India was first detected in West Bengal in 1983. Since then the number of states affected by this catastrophe has been on a steady rise. Today, arsenic contamination of groundwater has been discovered in Jharkhand, Bihar, Uttar Pradesh, Assam, Manipur, and Rajnandgaon village (Chhattisgarh). Experts even suspect that the north-eastern states of India also have a very strong probability of spewing arsenic contaminated groundwater. Therefore, one may conclude that the true extent of the crisis is yet to be ascertained and understood, and we strongly feel that scientific enquiry needs to continue. Confusion regarding proper management of the problem is still on the rise. Whether a piped supply of surface water is the most feasible solution to the problem is still an open question. Much of this is due to the fact that not much holistic assessment in the context of climate change impact at various time intervals on this subject is available. Moreover, questions regarding source security, security of distribution network, besides those related to usual cost recovery mechanisms still loom large. The probable set of solutions for the crisis, over the years, has gained considerable amount of bulk. These include, among others, suggestions to withdraw water from deeper aquifers, benchmarking of arsenic remediation technologies, and technology deployment to provide long-term secure sustainable water supply through source diversification. In short, finding solutions to the crisis has become quite a complex case in itself. This book provides detailed knowledge and information on the pros and cons of each alternative intervention and institutional arrangement towards arsenic mitigation that has been tried out in West Bengal with the hope that learning from the past will help in avoiding future mistakes in the next endeavours for solution.

Over three decades have passed since the initial detection of arsenic contamination of groundwater in West Bengal. Millions of litres of contaminated groundwater have been withdrawn and consumed during this period; several millions have been poisoned, slowly, one gulp at a time; several thousands have been affected by arsenic-induced illnesses; and several hundred, at least, have lost their lives.

It is not that there have been no mitigation efforts during this period. In fact, on the contrary, once the initial phase of bewildered disregard of the crisis, which at times seemed like bordering on naivety, was over, all empowered stakeholders jumped into the fray to help the millions of hapless affected. Funds came pouring in. Technologies were transferred, and also some were indigenously developed. Filtration mechanisms of different scales were commissioned. And surely arsenic was filtered out of groundwater. Success was achieved.

But the crisis persisted. And the need to have this book out in the public domain kept growing. And the need for it to be read matched pace.

The band of arsenic-defeaters, compelled by an unquenchable zeal to remove arsenic from groundwater, moved on to the next location. Like knights on a crusade they bludgeoned ahead and never looked back. In hindsight, had they looked back, they would have seen the trail of failures they were leaving behind. Our goal is to document these 'learning by doing' to help in the future decision-making process. Arsenic removal units converted into storage containers for cereals and grains; water posts converted into tying posts for domesticated animals; arsenic removal plants running for years without testing of water quality; or better still—plants where media has not been changed since installation: all these are but a reflection of this massive collective failure which can be summed up as the arsenic mitigation effort.

The idea of getting the field level realities documented and communicated to various stakeholder groups requires a skill set consisting of multiple expertises. This common need brought all of us, the authors, together. What began as the lead author's search for an interesting research topic for pursuing Ph.D. found in the second author the right guidance due to her already published research-based works on the subject. The third author found the first two while he was conceiving the idea of making a documentary film on the problem to report the plight to a much larger audience. That is how despite varied institutional affiliations and physical distances the research gap and thirst for knowledge dissemination led to this team building for writing this book volume.

The book is an effort at looking back to help the onward future journey. To sum up the journey so far, the authors have met many of the crusaders who have strived hard to solve the crisis. Because essentially the story of arsenic mitigation of groundwater is the story of people—of these crusaders who took it upon themselves to report, quantify and solve the crisis—of those who have suffered unknowingly and with the hope that everything will be alright—and those who continue to do so harbouring the same hope. Personal stories are testimonies from reality. This strengthens the scientific arguments of this work. "Good science is good observation."—Dr. Max Patel famously explained to Jake, the protagonist of the 2009 Hollywood blockbuster Avatar. For starters, we accept that referring to Hollywood for justification while writing a book on one of the gravest socio-economic crises is a serious travesty. However this is also the singular quality which, according to us, will force our book to make people think. The ‘From the field’ sections in the book simply put into perspective the socio-economic reality. It is, in other words, the ‘good observation’, or glimpses to it, that is fundamental towards the eventual success of any technology. In each chapter, this section will be a mirror to the implications of the remedial or mitigation measures undertaken by the crisis managers. These narrative sections mark the genesis of ‘good science’ or in other cases, great solutions that can truly alter lives, of millions.

The authors feel that this book will provide all with the much needed opportunity to contemplate the mitigation efforts till date. The entire mitigation exercise of arsenic contamination of groundwater crisis, at least in India, has been dealt with by key stakeholders as a scientific or technological crisis. Find the appropriate technology and voila, the problem is solved! But after almost 25 years of the identification and recognition of the crisis and subsequent discoveries of numerous technologies and even more mitigation efforts, the problem persists.

The authors also hope that as a by-product of this pondering, we will be inspired to de-learn and subsequently figure out a new transformative way ahead. After all, drinking water is a fundamental human right. Any right is not devoid of responsibility that needs to be addressed, effectively, consistently and sustainably. While on the one hand we stand by the fundamental right to water for all, we also firmly believe that another fundamental question of ‘burden sharing’ needs to be answered before we find a sustainable solution to the crisis: what is the cost of mitigation and who pays for the price of mitigation and non-mitigation?

We take this opportunity to express our sincerest gratitude to our institutions for allowing us to take up the assignment and complete the work successfully. During the course of this detailed study we came in touch with a number of domain experts from diverse fields and in turn we got enriched in various forms. We may inadvertently miss out some names, but still taking the risk to name them—Dr. Alakendu Ghosh, PG hospital; Dr. Alan Smith, University of California, Berkeley; Shri Animesh Bhattacharya, Assistant Chief Engineer, Public Health Engineering Department, Government of West Bengal; Mr. Arumugam Kalimuthu, Country Director—India, Water for People; Dr. Arunabha Majumder, School of Water Resources Engineering; Dr. Ashok Gadgil, Division Director, and Faculty Senior Scientist, Environmental Energy Technologies Division, LBNL, Andrew and Virginia Rudd Family Foundation, Distinguished Chair of Safe Water and Sanitation, Professor of Civil and Environmental Engineering, University of California, Berkeley; Dr. D.N. Guha Mazumder, DNGM Research Foundation; Dr. Dipankar Chakraborti, Director (Research), School of Environmental Studies, Jadavpur University; Shri Debdoot Ghosh Thakur, Chief Reporter, Anandabazar Patrika; Shri G.C. Pati, Regional Director, CGWB—Eastern Region; Dr. G. J. Samathanam, Advisor and Head, Technology Development and Transfer, Department of Science and Technology; Dr. K.C. Saha; Professor K.J. Nath, Chairman, Arsenic Task Force; Dr. Kshipra Misra, Additional Director/Scientist—F and Head, Department of Biochemical Sciences, Defence Institute of Physiology and Allied Sciences; Dr. Meera Smith, Project Well; Dr. N.C. Ghosh, Scientist—F and Head, Ground Water Hydrology Division, NIH; Shri Nirmal Pattanayak, Lok

Shiksha Parishad, Ramakrishna Mission Ashrama, Narendrapur; Shri P.O. Basu, Secretary, Save the Environment; Dr. P. Sikdar, Professor, Department of Environment Management, IISWBM; Dr. Pratap Chakraborty, GSI; Dr. Rajendra Singh, Chairman, Tarun Bharat Sangh; Dr. S.P. Sinha Roy, Member, Arsenic Task force, West Bengal; Shri Sandip Chatterjee, Engineer in Chief, Public Health Engineering Department, Government of West Bengal; Dr. Santanu Bhattacharjee, National Metallurgical Laboratory; Shri Saradindu Chakrabarti, Associate Professor and Co-ordinator, Department of Applied Geology and Environmental, Systems Management, Laboratory of Environmental Systems Management, Presidency University; Shri Shyamnarayan Dave, WASH Specialist, UNICEF Field Office for West Bengal; Dr. Sibdas Bandyopadhyay, Chief Scientist and Head, Ceramic Membrane Division, Central Glass and Ceramic Research Institute; Dr. Supradip Sarkar, CPI, NAIP-04 Sub-project, Directorate of Research, BCKV; Dr. Susan Amrose, University of California, Berkeley; Shri T.M. Vijay Bhaskar, Joint Secretary (Water), Ministry of Drinking Water and Sanitation, Rajiv Gandhi National Drinking Water Mission, Government of India; Dr. Tarak Nath Pal, Senior Geologist, Central Petrological Laboratories, GSI. There are many more who have inspired us to embark on this remarkable journey to bring to fore the inside story of arsenic crisis.

We are grateful to students (Shri Asish Majhi, Shri Rahul Sinha, Shri Subhash Ghosh, Shri Sudip Mondal, Shri Sukalyan Upadhay, and Shri Taimur Bin Kashim Khan) of Kandi Raj College (Murshidabad, West Bengal) for helping in the surveys conducted in Murshidabad district funded by University Grants Commission. We are grateful to all those (Shri Achirangshu Acharyya, Shri Abhinanda Deb, Shri Arunagsu Chattaraj, Shri Bikram Chatterjee, Shri Raman Khaddaria, Shrimati Sohini Sahu, and Shri Suman Dutta) who helped in conducting the survey and subsequent data analysis for the North 24 Parganas study funded by South Asian Network for Development and Environmental Economics (SANDEE).

What has driven us and kept us together is the genuine wish to advance the knowledge and to promote the idea of learning from the past for defining the way forward on the subject and deep concern for human welfare where situations can and, in the case of arsenic contamination of groundwater have gone beyond control —spiralling into becoming a complex crisis. So this book is the authors' effort to portray the unbiased reality based on learning from the past; bring to fore existing solution spaces; and unequivocally emphasise the dire need for social placement of technology, policy and action towards sustainable solution to effectively solve the groundwater arsenic crisis.

Contents

About the Authors

Dr. Abhijit Das is currently Assistant Professor in Economics at Vijaygarh Jyotish Ray College in Kolkata, West Bengal (India). He has completed his doctoral degree on socio-economic aspect of groundwater arsenic pollution in West Bengal from Jadavpur University. He has been researching extensively over the past one decade in various villages of rural West Bengal to understand the social, economic and political issues that drive lack of access to safe drinking water. His understanding of the problem is based on rigorous research and extensive interaction with the rural communities, government and non-governmental service providers. His field research has been primarily funded by the University Grants Commission of Government of India. He authored a book in Bengali 'Jale Arsenic-Prasanga Pashimbangya' in association with Professor Joyashree Roy. More than 15 research articles have already been published in different national and international journals as well as book chapters. He also presented his research findings on arsenic problems in West Bengal in different international and national seminars and conferences.

Dr. Joyashree Roy is currently Professor of Economics, coordinates the Global Change Programme and also directs Ryoichi Sasakawa Young Leaders Fellowship Fund Project at Jadavpur University. She was on the 2007 Nobel Peace Prize winning panel—IPCC (Intergovernmental panel on climate change). She has been involved in preparation of the Stern Review Report, Global Energy Assessment, and many other national and global reports. She has published more than 80 peer-reviewed articles.

Sayantan Chakrabarti is a development communication practitioner who has worked on issues as diverse as AIDS to heritage conservation to arsenic contamination of groundwater. In the process he has created gripping documentaries for UNICEF, West Bengal Board of Secondary Education, Jamia Millia Islamia, Good Morning Channel (USA), IL&FS IWMUSL and PSBT amongst others.

About the Book

Issues linking technology innovation in providing arsenic safe drinking water in rural areas, and breaking the social barriers for technology uptake are now at the centre of discussions regarding developmental strategies through technology deployment especially in the context of developing countries. This book contributes by depicting real case studies to show the plight, sufferings and failed technology deployment strategies implemented over the decades in the State of West Bengal, the most arsenic exposed state in India. The book also provides rigorous socio-economic analysis of the situation, but in a very reader friendly language without compromising on the scientific facts.

One of the most complicated problems in arsenic exposed zones is to provide in a sustainable manner access to reliable arsenic safe water to the affected population. For almost two decades government and national water distribution institutions in collaboration with multilateral funding agencies tried hard to pump in money, push technology collected through global tenders, involve NGOs to look after the people connect system, and involve community to look after their own welfare, but unfortunately the real progress has been negligible. We are, honestly, at almost the same point where we were three decades back. This book is an outcome of very intensive fieldwork undertaken by the authors using conceptual framework of most advanced literature on environmental economics and consumer behaviour. The book presents not only the framework but also estimates based on primary field level data. It also collates secondary official source-based data into a very valuable collection. In short, this book provides a very important database that collates the complete story, so far, of existing endeavours as well as past attempts towards mitigation of the arsenic contamination of groundwater and reasons for their failure. The novelty of the book lies in the fact that for the first time for West Bengal a comprehensive documentation on arsenic contaminated drinking water use, remedies, outcome and way forward is prepared with a detailed presentation of the methodology, in English, to communicate with the larger audience. We did publish a book in Bengali keeping in mind the readers of Bengal, but were flooded with requests to narrate the saga of the ordeal in English. This book is an effort to honour those very demands. The book also includes analyses of factors affecting household participation in the mitigation process. The book will have relevance not only for India but also for Bangladesh, which is known for its disastrous impact of arsenic contaminated groundwater use for drinking purposes. To sum up, the book is an important contribution in the current discourse on how to go about solving the problem.

Acronyms

AA	Activated Alumina
AAN	Asian Arsenic Network
AAS	Atomic Absorption Spectrometer/Spectrometry
ADB	Asian Development Bank
AE	Averting Expenditure
AFS	Atomic Fluorescence Spectroscopy
AIIH&PH	All India Institute of Hygiene and Public Health
AMCL	Acceptable Maximum Contamination Level
AMT	Arsenic Mitigation Technology
ARP	Arsenic Removal Plant
ARWSP	Accelerated Rural Water Supply Programme
As	Arsenic
ASD	Arsenical Dermatosis
ATF	Arsenic Task Force
ATU	Arsenic Treatment Unit
ATU-HP	Arsenic Treatment Unit-Hand Pump
BAL	British anti- Lewisite (or Dimercaprol)
BARC	Bhaba Atomic Research Centre
BAU	Business As Usual
B.E. College	Bengal Engineering College
BE	Budget Estimate
BIS	Bureau of Indian Standards
CAZRI	Central Arid Zone Research Institute
CCDU	Community Capacity Development Unit
CGCRI	Central Glass and Ceramic Research Institution
CGWB	Central Ground Water Board
CMRI	Central Mechanical Research Institute
CSIR	Council of Scientific and Industrial Research
CSMCRI	Central Salt and Marine Chemicals Research Institute
CSME	Centre for Study of Man and Environment
CVM	Contingent Valuation Method

DCH	Dhaka Community Hospital
DDWS	Department of Drinking Water Supply
DF	Domestic Filter
DHS	Directorate of Health Services
DMPS	Dimercapto Propane Sulphonate
DMSA	Dimercapto Succinic Acid
DWQC&SP	Drinking Water Quality Control and Surveillance Programme
EC	European Commission
ECAR	Electro Chemical Arsenic Remediation
EPA	Environmental Protection Agency (United States)
ER	Eastern Region
FC	Fully Covered
GCP	Global Change Programme
GDWQ	Guidelines for Drinking Water Quality
GMB plain	Ganga–Meghna–Brahmaputra plain
GOI	Government of India
GOWB	Government of West Bengal
GP	Gram Panchayat
GSI	Geological Survey of India
Hhs	Households
HIG	High Income Group
HRD	Human Resource Development
HRF	Horizontal Roughing Filter
IARC	International Agency for Research on Cancer
ICEF	India Canada Environment Fund
IEC	Information, Education and Communication
IEMS	Institute of Environmental Management and Studies
IIAWQM	International Institute of Arsenic and Water Quality Management
IPCC	Inter-government Panel for Climatic Change
IPGME&R	Department of Gastroenterology, Institute of Post Graduate Medicine Education and Research
IQ	Intelligence Quotient
ITRI	Industrial & Toxicological Research Institute
JPOA	Joint Plan of Action
JU	Jadavpur University
KMD	Kolkata Municipality Development
KU	Kalyani University
Lao PDR	Lao People's Democratic Republic
LIG	Low Income Group
lpcd	Litre per capita per day
Lt	Litre
MDG	Millennium Development Goal
MIG	Middle Income Group
MNP	Minimum Needs Programme

MPLAD Fund	MP Land Area Development Fund
MWTP	Marginal Willingness to Pay
NC	Not Covered
NDWM	National Drinking Water Mission
NEERI	National Environmental Engineering Research Institute
NFHS	National Family Health Survey
NGO	Nongovernmental Organisation
NIH	National Institute of Homeopathy
NIH	National Institute of Hydrology
NML	National Metallurgical Laboratory
NMRL	Naval Material Research Laboratory
NRDWQM & SP	National Rural Drinking Water Quality and Monitoring and Surveillance Programme
NSSO	National Sample Survey Organization
O&M	Operation and Maintenance
PC	Partially Covered
PHC	Primary Health Centre
PHED	Public Health Engineering Directorate
PMB	Padma–Meghna–Brahmaputra
PMGY	Pradhan Mantri Gramodyog Yojana
PMIND	Prime Minister's Independence Day Announcement
PPB	Parts Per Billion
PPM	Parts Per Million
PRI	Panchayati Raj Institution
PWSS	Pipe Water Supply Scheme
R&D	Research and Development
RGNDWM	Rajib Gandhi National Drinking Water Mission
RKMLSP	Ram Krishna Mission Lok Shiksha Parishad
ROHC	Regional Occupational Health Centre
RWS	Rural Water Supply
SAR	Subterranean Arsenic Removal
SFR	School of Fundamental Research
SM	Sanitary Marts
SOES	School of Environmental Studies
SRP	Sectoral Reform Programme
SSF	Slow Sand Filtration
SSKM Hospital	Seth Sukhlal Karnani Memorial Hospital
STM	School of Tropical Medicine
SWID	State Water Investigation Directorate
TP	Technology Park
TSC	Total Sanitation Campaign
UC	Un-Covered
UN	United Nations
UNCED	United Nations Commission on Environment and Development

UNDP	United Nations Development Programme
UNICEF	United Nations Children's Fund
UP	Uttar Pradesh
VWSC	Village Water and Sanitation Committee
WB	West Bengal
WFP	Water for People
WHO	World Health Organization
WTP	Willingness to Pay
ZP	Zilla Parishad

Units of Measurement

Cap^{-1}	Per Capita
cm	Centimetre
Cr	Crores
Gpcd	Gallon per capita per day
kg	Kilogram
kl	Kilo Litre
kld	Kilolitre/daily
km	Kilometre
km^3	Cubic Kilometre
Lpcd	Litre per capita per day
Lt	Litre
m	Meter
mg/l	Milligrams/litre
m^3	Cubic Meter
Mgd	Million gallon daily
mld	Millilitre daily
mm	Millimetre
mg	Milligram
Rs	Rupees
Yr^{-1}	Per Year
μg	Microgram
μg/L	Micrograms/Litre

Conversion Formulae

1 gallon	= 4.54 litres
1KL	= 1000 litres
1 cubic meter	= 1000 litres
1 mg/L	= 1 ppm = 1000 μg/L
1 μg/L	= 1 ppb

The Plight

Abstract To understand the true impact of the arsenic contamination of groundwater problem in West Bengal, it is imperative that one comprehends its micro and macro dimensions.What has been the extent of the problem and how the sufferings of affected people living in arsenic exposed and infested villages of West Bengal have remained unnoticed and unattended is narrated using information collated during our personal and group interview sessions. The names of respondents have not been changed according to their wishes.Moreover, we adopted a methodology that consisted of both the quantitative and qualitative analysing techniques in the form of review of literature; official statistics; and, of course, in-depth case studies.

What does an animal in a zoo feel after a day's work of being displayed in its cage?

Mr. Mahafijur Rehman's reaction to our visit brought up this question in our mind!

Mr. Rehman is from Panjrapara village, Khoiramari gram panchayat (GP), Jalangi block, Murshidabad district, West Bengal, India. A little perspective on Murshidabad may help you understand the plight of Mr. Rehman, better.

In 2006, the Ministry of Panchayati Raj, Government of India, named Murshidabad as one of the country's 250 most backward districts[1] (out of a total of 640). Almost 63.7 % of its population is Muslim,[2] the highest in West Bengal. Almost 20.9 % of its people are living under the poverty line.[3] According to the District Level Household and Facility Survey, 2007–08, only 15.9 % villages in Murshidabad have access to qualified doctors and 43.2 % have access to any kind of government health facility!

To make matters worse, groundwater, the principal source of water for drinking and other forms of consumption, in Jalangi, the block where Mr. Rehman resides, is

---

[1]Backward Regions Grant Fund Programme Guidelines: prepared and circulated by Ministry of Panchayati Raj, Government of India, Release Date: 15.01.2007.

[2]http://www.censusindia.gov.in/Census_Data_2001/Census_Data_Online/Social_and_cultural/Religion.aspx accessed on 12.02.13.

[3]International Institute for Population Sciences (IIPS), 2010. District Level Household and Facility Survey (DLHS-3), 2007-08: India. West Bengal: Mumbai: IIPS.

A. Das et al., *Socio-Economic Analysis of Arsenic Contamination of Groundwater in West Bengal*, India Studies in Business and Economics, DOI 10.1007/978-981-10-0682-1_1

severely contaminated. According to the Public Health Engineering Department (PHED), Government of West Bengal (GoWB) report, all the GPs of Jalangi block are arsenic affected. According to that report, i.e. Census 2001[4] by PHED, GoWB, 216 000 people residing in Jalangi who spread over some 210.63 km^2 are declared as population at risk due to the arsenic crisis. It translates to, in strictly statistical terms, the probability that many residents of this region will have arsenic-induced illnesses and eventually may even die due to those.

Mahafijur is one of the many who have been affected by the poison, irreversibly. As a result, he has been exhibited to and searched for by academics, journalists, doctors and all individuals involved with the arsenic crisis and its mitigation measures in some form or the other for more than a decade now. His images have adorned the pages of important international journals that he has never heard of. His heart-rending story has drawn sympathy. Probably, they have even compelled a flow of funds to mitigate the crisis. But, his situation has not changed.

During our visit his anger, fuelled by years of frustration, simply seethed over. He asked, "We don't think of arsenic anymore. What will change by thinking about it? What difference are you going to make?" We had no appropriate answer.

Mr. Alauddin Sarkar, his neighbour, has had a similar fate. He has distinct spots all over his body. He cloaks those with clothes. But, those on the palms and soles need regular laceration. But, even this painful exercise does not give him much relief. He cannot even walk barefoot, let alone the possibility of working. Mr. Sarkar considers the tube well just outside his premises, once a distinct landmark for his house which also turned the location of his house a rather enviable one, his personal enemy. He still draws water from the same, rather unwillingly. He knows the water it spews is essentially poisoned. Over the years, his knowledge about arsenic and probable mitigation approaches has been enriched. He has been informed that water from a dug well is arsenic safe. But, ironically, he has no means of acting on the basis of this information. In fact, he can barely make the two ends meet. He has resigned to his fate. But, he hopes that safe drinking water will be available for future generations. His hope has been resting on the government to make this dream come true. So far, it has not reached fruition. His family has remained in the category of 'Population at Risk' according to the previously referred to PHED, GoWB report, even a dozen years after its publication. The major grouse for the affected population, all through these years, has been the obvious unavailability of safe drinking water. The other bone of contention amongst all stakeholders and experts has been about the 'at Risk' part of the statement. It might be important at this juncture to comprehend, for once and all, the competing definitions and standards about 'at Risk.'[5]

[4]Refer to Annexure 1, Table A.1.

[5]Later in a separate chapter we report statewide numbers just to highlight the severe discrepancies even in the mere numbers as churned out by government and independent surveys and research organisations.

From the Field 1

Mr. Rustam Sheikh was a strong man. Well, his able body was his only asset to leverage him and his family out of the abject poverty that he was born into. He used to work as a daily wage labourer mainly by tilling other people farms. His other possession: a piece of land, the size of a handkerchief, where he cultivated the occasional vegetables and pulses to put something on the plate for his family. It can be safely stated that Rustam managed to meet two ends meet.

About a dozen years ago, it all slipped out of his hands, almost literally. Arsenic was discovered in all the tube wells of his locality. His ability to toil in the sun had been on a steady decline for quite some time before that. There also were few deaths in the locality. But, till that date no one had considered water to be the culprit.

Around the same time, a cancerous wound appeared on the index finger of his right hand. Mr. Sheikh lent some money and visited a doctor, initially at Murshidabad and subsequently at Kolkata. He managed to get a letter from the local MLA which eased him into the Kolkata hospital with relative ease. After all, Kolkata at that point of time had several government hospitals offering medical assistance to individuals affected by arsenic-induced diseases. Doctors at the Kolkata hospital operated on him. He came back home with a finger missing.

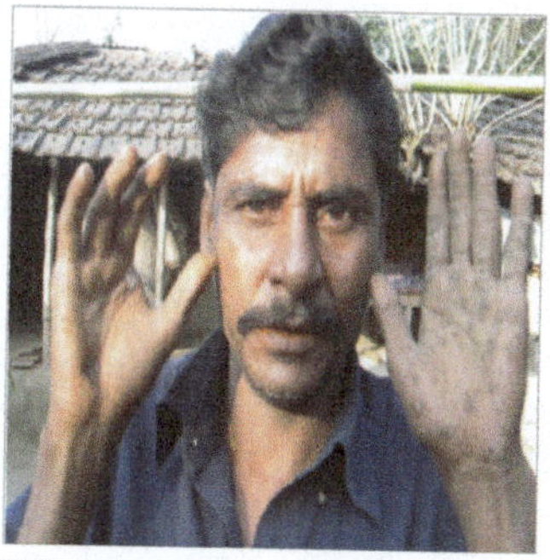

Rustam during our first visit in 2009

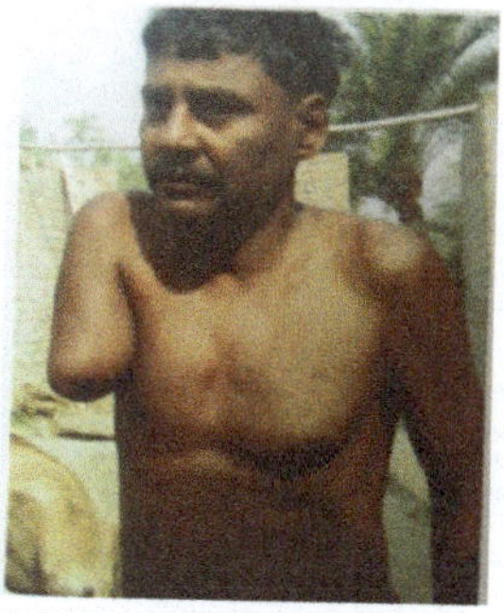

The last time we met Rustam during our visit in 2013

However, the cancer spread, and the entire process needed to be repeated. Again, he visited the MLA for the letter, borrowed some more money, and embarked on a journey towards Kolkata. During this visit, doctors decided that the hand cannot be saved. So, he underwent another operation and amputated his right hand up to the wrist. But, the advent of cancer in his body continued, unabated. Yet again it appeared. This time, the hand was severed above the elbow, leaving a stump like entity which at best hints at an erstwhile limb.

Gradually, Mr. Sheikh became almost invalid. Severe skin lesions developed all over his body. He could not go out in the sun to work anymore. His little piece of land started serving only as a grazing field for wandering cattle. Worse, whatever remained of his right hand needed to be operated again. This time, the doctors felt they needed to take it out right up to the shoulder in an effort to prevent further spread of cancer in his body. However, going by past experiences, there was a strong possibility that this last ditch effort may also prove to be inadequate.

He had been advised resettlement many a times but had always refused against it. He said, "Fleeing the crisis is not a solution. Why would I have to leave my birthplace in order to survive?" His son, however has moved out. The boy works as a daily wage worker in a different state. It is of his income that Mr. Sheikh tried to survive with a semblance of dignity. After all, he was one of the most prominent victims of arsenic-induced illnesses and had no dearth of visitors. The only thing which had kept pace with the rampaging cancer had been his severe apathy towards visitors.

Sometime in the month of August, 2014, Rustam's fight ended. But, the struggle continues for the rest of his family. Today, the steel which once resonated in Rustam's voice finds its expression in his wife's when she asks visitors to leave her family alone.

How Much Is Too Much?

Abstract The fact that arsenic is poisonous is well known. Complete avoidance is the best preventive action. However, this remains an implausible situation in arsenic affected areas of West Bengal. The obvious question that arises then is what is the safe or permissible limit of arsenic in drinking water. To find out the answer to this critical question, we reviewed literature that spells out the diverse and contrasting safe limits prescribed by various international nodal authorities as well as national governments. These variations are partly due to the fact that political and administrative policies are, at least in the case of West Bengal, if not India, limited by the lack of dedicated research and specific information and know-how. Moreover, apart from the fact that the policies are not always devised based on scientific information they are also, at times, heavily weighed down by several other operation-related factors. However, this adherence to inefficient safe levels brings forth several real-life implications for those suffering.

Arsenic has been clinically proven to cause adverse health effects even at very low concentrations.[1, 2, 3, 4, 5, 6] At the same time, it ranks 20th in the list of naturally occurring elements in earth's crust and hence its presence in water cannot be completely avoided.[7] Therefore, the standard limit for arsenic in drinking water is set accepting certain risks that do not result in any adverse health effects over the human lifetime of water consumption. The World Health Organisation's According to (WHO) most recent guideline values for a standard person (60 kg of body weight with a lifespan of 70 years) drinking 2 L of water per day is 10 μg/L. In other words, 10 μg/L of arsenic in drinking water will not mortally affect an individual weighing 60 kg over a lifetime of 70 years of drinking 2 L of water daily. Just to put

[1]Navas-Acien et al. (2008).
[2]Moon et al. (2013).
[3]Dangleben et al. (2013).
[4]Gilbert-Diamond et al. (2013).
[5]García-Esquinas et al. (2013).
[6]Mostafa and Cherry (2013).
[7]Refer to Table A.4.

A. Das et al., *Socio-Economic Analysis of Arsenic Contamination of Groundwater in West Bengal*, India Studies in Business and Economics,
DOI 10.1007/978-981-10-0682-1_2

into perspective, to arrive at this value WHO has deliberated for over four and a half decades.[8]

The guideline thus arrived at, however, is merely suggestive. Countries[9] have adopted WHO guidelines from different eras to suit their convenience and abilities. Interestingly, there are countries willing to push the permissible limit to even lower levels, e.g., Australia has adopted 7 μg/L as its standard. Incidentally, in India, the maximum permissible limit is fixed at 50 μg/L[10]; a figure arrived at by WHO in 1963! According to one section of the stakeholders, this level is simply not good enough. The source of this displeasure is principally two reports; one by National Research Council of America, 2001 and the other by Science Journal, 2002. These reports stated that there is a possibility of 13 deaths due to cancer in every 1000 individuals if they consume 1 L of water containing 50 μg of arsenic, regularly, over a lifetime. Manifestations of this threat get even uglier due to the poor level of nutrition,[11, 12] coupled with inadequate healthcare facility for many of those exposed, especially in developing countries like India. Based on these observations, the Bureau of Indian Standards (BIS) recommended in September 2003, that 10 μg/L should be adopted as the standard of arsenic in drinking water. Shortly after this announcement, a publication (Smith and Hira Smith 2004) suggested raising it to 50 μg/L for India and other developing countries due to current resource limitations and the likely slowing of short-term solutions imposed by this stricter standard. In a somewhat kneejerk reaction, partly influenced by this publication, the Indian standard was quickly reinstated to 50 μg/L. This decision did not go down well with one section of the stakeholders who argued that the potential health effects of arsenic exposure were too severe even at concentrations below 50 μg/L and that concerned authorities should strive to overcome resource limitations. In an attempted compromise, BIS has recently published revised standards for arsenic in drinking water, listing 10 μg/L as the desirable level but 50 μg/L as the legally enforceable standard if alternative sources are unavailable (Chakraborti et al. 2011).

At this juncture, questions may arise, naturally, about the justification of lowering of the safe level. To be honest, we could find none. It cannot be justified under neither scientific nor any human welfare rationale that why should Indians be exposed to higher arsenic concentration levels. The only beneficiary out of this play of words is, probably, the administrative organisations. The lowered level allows them some breathing space and in effect justifies their inefficiency. For example, if WHO standard is to be adopted, then 56 % of end user-accessed water service points (tube wells) in West Bengal are arsenic unsafe. On the other hand, adherence to the national legally enforceable standard reduces the unsafe percentage by less than half, i.e. 26 %. There are ample dermatological surveys (e.g. Saha and

---

[8]Refer to Table A.2.

[9]Refer to Table A.3.

[10]Indian Standards for Drinking Water, second revision of IS 10500 (2004).

[11]Maharjan et al. (2007).

[12]Mitra et al. (2002).

Chakraborti 2001) to prove that exposed population not always exhibit similar manifestations in spite of having comparable arsenic intake. But, clinical examination of their hair, nail and urine reveals high concentration of arsenic! May be several of these villagers are subclinically affected, and it is just a matter of time before an arsenic-induced epidemic explodes on the public sphere.

All this while, at the ground level, safe water infrastructure development has, at best, made crawling progress. Resultantly, the 'Population at Risk', like those we visited in Jalangi, merely survives under the shadow of imminent death. Nobody knows how many of them will be diagnosed with cancer and die soon. No wonder, they seem a bit bitter towards visitors. But, maybe they still harbour some hope that things will get right if they participate in yet another survey, and eventually choose to talk to yet another team of researchers.

From the Field 2

Sometime in the year 2002, the villagers of Plachimada, a small village of Palakkad district, Kerala united in their struggle against the behemoth of an MNC, Coca Cola. They accused the global giant of stealing and polluting their water. The source of the trouble was the Coca Cola plant in the village and the untreated and highly toxic effluents it was dumping into the groundwater.

Around the same time, the villagers of Harekrishnapur, Khairamari GP, Jalangi block, Murshidabad started getting very angry too. Their water, like in Plachimada, was found to contain alarming levels of poison. The poison content, on an average, which in their case was Arsenic, was about five times safe national standard. But, that is where the similarity of their predicament to those in Plachimada ends. The residents of Harekhrishnapur did not know whom they should unite against in protest. They do not know even now!

For the last decade or so, those living in Harekrishnapur have experienced fleeting moments of hope amidst continual feelings of despair. The list of ill and dead keeps growing by the day. So is the extent of anger. The hapless people are confused about who can bring an end to their misery, and therefore have believed at every promise of solution made to them, gullibly. Resultantly, today they are sceptic and utterly bitter. In their words, 'only death has kept its promise.' They hate having visitors. They dread promises.

When we visited Harekrishnapur, despite being named after Lord Krishna, the God of love and affinity, we found its residents rather hostile towards us. They ordered us to walk away. It was apparent that the villagers have had enough of sharing of their ordeal with information-hungry researchers and media personnel and tired of their false sympathies which inadvertently resulted in heartbreaks. When at last they agreed to speak with us the conversation began with a barrage of questions. 'Do you know how many families have lost their near and dear ones? How many people have had their organs amputated? How many people are suffering silently? How many children have lost their childhood, and education, and have ended up as child labourers?' they asked. 'Do you know how difficult is to find a groom for our daughters and sisters?' they continued. We, obviously, had no answer to offer. 'There are several dug-wells in the village. They are all dry and all we have are skeletons of defunct Arsenic Treatment Units! This is a village where people live, not a laboratory dedicated to the cause of advancement of science!' they concluded. This time they expected no answer or for that matter reaction from us.

Our mind could not but refer back to the Plachimada example. The international media was present, so were water crusaders from all over the world. It seems Coca Cola, being a private company can be held accountable. However, the government, at least in context of arsenic contamination of groundwater in Harekrishnapur, is not accountable to the public. The residents of Harekrishnapur are not certain whom to appeal to, to solve their crisis. However, there is the certainty that death will arrive, sooner rather than later.

For us, it was an ignominy never felt before.

Right to Water

Abstract Water supply is a state's responsibility. Allocation of safe water, or in the case of arsenic-affected regions of West Bengal having the lack of it, therefore has been at the centre of the crisis. The 'Right to water' as mentioned by United Nations is yet to be realised for the common man. In arsenic-infested regions the problem takes on even ominous implications. Moreover, with the ever burgeoning population and discovery of newer contaminants the water stress of the nation is growing by the hour. The book adheres to literature review and empirical evidences to elucidate the situation.

U.N. General Comment number 15 on the Right to Water was made public in 2002. It stated, "The human right to water entitles everyone to sufficient, safe, acceptable, physically accessible, and affordable water for personal and domestic use". It is now proven that making 'safe' water accessible at the end user point is an essential service that is necessary for maintaining satisfactory public health standard and the optimal level of human well being.[1] In the context of India and other developing countries, water borne diseases dampen economic growth and worsen poverty status. In fact, deteriorating water quality has been identified as a major hurdle towards realisation of India's sustainable development agenda (Roy 2008).

All said and done, there is no doubt that the common man is yet to realise his 'Right to Water'. How else can you justify the death of almost 2 million children each year for want of safe water and adequate sanitation? In fact, unsafe water accounts for the second highest number of child deaths per year globally. In 2006[2] alone, 443 millions of school days were lost due to water related illnesses and one in five people living in the developing world, in other words, some 1.1 billion people had no access to clean water. Unfortunately, this ordeal is even harsher on women, the traditional water provider to families, in developing countries.

[1]Roy (2008).
[2]UNDP (2006).

A. Das et al., *Socio-Economic Analysis of Arsenic Contamination of Groundwater in West Bengal*, India Studies in Business and Economics, DOI 10.1007/978-981-10-0682-1_3

Statistically speaking, tens of millions of women and children spend eight or more hours, daily, fetching water (Ghosh et al. 1995).

The world is becoming 'water stressed,[3] fast. Today, about 700 million people in 43 countries live below the water-stress threshold. By 2025 that figure will reach 3 billion, as water stress intensifies in several Asian countries and Sub-Saharan Africa (UNDP 2006). The case of India is no different. In 1996, India's per capita availability of water was 2200 m^3 (Water Aid 2008). The comparative figure for 2025, approximately, will be a mere 1600 m^3. We are simply exhausting more than what even the mighty monsoon is able to replenish.

The principal culprit for this overexploitation is the decision towards limitless deployment of mechanised and as well as handpump attached tube wells. The number of handpumps rose from less than a million in 1960 to an estimated 19 million in 2000 (Shah 2007) mainly due to encouragement from several national as well as international agencies. The groundwater use, fuelled by rise in population, food production and rapid urbanisation, in cumulative terms soared from around 10–12 km^3 before 1950 to 240–260 km^3 in 2000 (Shah et al. 2005). However, the drinking water component in this rise was very small and without any significant implication on withdrawal of groundwater. This exploitation, however, brought in problems related to not only depleting quantity, but also of contaminants not dealt with before. Primary among those were fluoride and arsenic. It is important to understand that principles and rules of groundwater withdrawal have been proven to have a clear bearing on the crisis. It is estimated that 60 million people in 20 endemic states are at risk due to excess fluoride (Doctor's Handbook 2005) and around 10 million people are at risk due to excess arsenic in groundwater (Water Aid 2008),[4, 5] in India.

From the Field 3

Uma Mondal of Harekrishnapur, like many of her neighbours, is an early riser. However, there is a specific purpose to her waking up at the crack of the dawn: fetching safe drinking water. The nearest safe water source, a deep tube-well, is approximately 5 km from her house.

Uma's household consists of seven members. The daily water requirement for her family is at least 30 L. That means, every morning Uma makes the arduous journey to the water source with two containers of considerable sizes. She takes a shortcut to save time. By shortcut, we mean a path that bears no resemblance to any metalled road. Rather it is an undulating surface conceived ingeniously across

---

[3]A country is said to be water stressed when the per capita availability of water drops below 1700 m^3/person/year (IPCC 2000).

[4]Refer to Table A.5.

[5]Refer to Table A.6.

dividers of agricultural fields. Commuting this path in good weather was a feat that is rather difficult to accomplish. However, she makes this trip daily. Thankfully, she usually has few of her neighbours keeping her company.

Uma, by nature who is jovial, now hates to host visitors. 'I have asked my sister to not visit me,' she says. The reason: more people mean greater consumption of water. That, in effect, means that Uma needs to embark on yet another water voyage.

Till recently, Uma ensured that sourcing of water does not become a burden on her adolescent daughter. However, since last monsoon, she has had to make an exception. Sometime in August, 2014, she had some visitors at house and was forced to go for a second water collection voyage. It had been raining heavily since the morning and the shortcut was in a really treacherous state. On her way back, she slipped and hurt her back, badly. Today she is forced to take her daughter alone. She is afraid to let her daughter make the trip alone. 'The times are bad and my daughter is pretty,' she sighs.

The Shades of Suffering

Abstract On the face of it, there have been several interventions by the state and its agencies to solve the crisis. But unfortunately, even after several years of planning the implementation has been, at best, sketchy. In this chapter, we put forth the various recommendations and plan objectives undertaken by the government to address the situation over the years. As a part of its intervention initiatives, the government has also set up substantial number of organisations and institutes partaking scientific research to device technological solutions. We also look closely at the roles played by these various organisations and institutes involved in the process of supplying safe drinking water by analysing empirical evidence.

In India, water in all its forms have been eulogised since the Vedic Ages. The Harappa civilisation had provisions for safe water supply and waste water disposal. Yet during 1961, only 1.9 % of rural India had access to water supply of any kind. The loss inflicted by this paucity was severe.

No wonder, safe water supply was one of the foremost priorities of the Government of India after gaining independence. In the First Five Year Plan (1951–56) there were allocations for safe drinking water supply to both urban and rural areas. Subsequent plans showed similar intentions. But the success they achieved on ground at best is debatable. The special provision for supply of potable water to the rural areas as part of the minimum needs programme (MNP) made in the Fifth Five Year Plan (1974–79) may be interpreted as a sign of failure of previous allocations to realise commendable successes. One thing is however undeniable, the policies have constantly been modified to keep up with the changes. In the seventh Five Year Plan (1985–1990) the factor of time bound delivery of safe drinking water was introduced. Whereas, the tenth Five Year Plan (2002–07) called for sustainability of safe water delivery. The approach paper of the eleventh Five Year Plan (2007–2012) calls for a comprehensive approach which encompasses individual health care, public health, sanitation, clean drinking water, access to food and knowledge about hygiene and feeding practice. It was envisaged to provide clean water for all by 2009 and ensure that there are no slip backs by the end of

A. Das et al., *Socio-Economic Analysis of Arsenic Contamination of Groundwater in West Bengal*, India Studies in Business and Economics, DOI 10.1007/978-981-10-0682-1_4

2012. Our independent research carried out in arsenic prone regions of Murshidabad, West Bengal, however concludes that the goal is yet to be realised.

Symbolising the commitment of the government to solve the safe water crisis are several institutional arrangements and technological interventions. A number of institutions have been engaged over time to review and solve the water quality problems in India. Technical institutions such as the Council of Scientific and Industrial Research (CSIR) Laboratories, Industrial and Toxicological Research Institute (ITRI), Lucknow; National Environmental Engineering Research Institute (NEERI), Nagpur; Central Salt and Marine Chemicals Research Institute (CSMCRI), Bhavnagar; Central Mechanical Research Institute, (CMRI), Durgapur; Defence Laboratory, Jodhpur; Central Arid Zone Research Institute (CAZRI), Jodhpur; All India Institute of Hygiene and Public Health (AIIH&PH), Kolkata; IITs; Regional Research Laboratories; Bhaba Atomic Research Centre (BARC), Mumbai; and National Institute of Communicable Diseases (NICD), Delhi have been effectively involved in the development of most appropriate technologies under various Research and Development (R&D) programmes sponsored by Department of Drinking Water Supply (DDWS) since the inception of the Technology Mission in 1986. These institutions have suggested number of technological options to tackle the water quality issues.

Considerable budgetary allocations followed to actualise many of these suggestions. International donors as well as state and central coffers pumped in funds for putting in place mitigation measures, both long and short term, almost intermittently throughout this period. Yet reliance on groundwater has not reduced and worse arsenic contamination of groundwater has been detected in henceforth unaffected regions. It has taken over previously unaffected sources and rendered water from newly sunk deep tube wells, one of the proposed and supposed solutions to the crisis unsuitable for consumption. This gradual discovery of affected regions and sources along with lack of knowledge about subsequent deterioration of quality can be attributed to the severe lack of surveillance of water quality. The detection of contaminants, at least in the case of arsenic, historically, has been a factor of chance rather than practice.

Take for example, the discovery of arsenic in groundwater in West Bengal in the 1980s. It was largely due to the inquisitiveness of Dr. K.C. Saha, the then Head of the Department of Dermatology of School of Tropical Medicine, Kolkata (STM). It all started when Dr. Saha confronted few rather perplexing cases in his outdoor section. The medical history of these patients in question stated Addison's disease. But Dr. Saha suspected otherwise. He had a hunch. May be the water they drink was the source of this peculiar disease. Based on this suspicion he ordered for a testing of the water they drank. The initial reaction to his recommendation was scornful disbelief. 'Arsenic in water?' ridiculed the person in charge of the laboratory. Once the results came, truth awed them all. Arsenic level in the water sample was found to be ten times more than the safe level. For the next few years, Dr. Saha travelled extensively to the villages from where the initial cases originated. Alone! He diagnosed patients, collected water samples for scrutiny, documented several patients, and sometime in 1984 reported in detail about the situation to the government of West Bengal. At that point in time, three villages of two blocks in two districts (North and South 24-Parganas) of West Bengal were affected allegedly.

The reportage was initially met with denial. But a few likeminded individuals took up the cause of arsenic victims. Foremost among those were Dr. D.N. Guha Mazumder, Head of Arsenic Clinic, Seth Sukhlal Karnani Memorial Hospital, Kolkata (SSKM) and Dr. Dipankar Chakraborti, Director of School of Environmental Studies, Jadavpur University (SoES). And the Pandora's Box was opened.

Arsenic was found in groundwater in many locations. The number of affected grew in tandem. And eventually, the statistics compelled the government out of the denial mode. Surveys by independent researchers still differed from those by government agencies. But the bone of contention this time around was the actual number of affected sources and individuals. This difference can be attributed to the disparity in coverage of their study areas. After all, the government's effort[1] to enumerate the affected regions and patients started much later than the private endeavours. It is now unanimously agreed that eight districts of West Bengal have unsafe levels of arsenic in ground water. These are Nadia, Murshidabad, Malda, North and South 24-Parganas, Howrah, Hooghly and Burdwan.

From the Field 4

Jahirul Islam from Gangarampur (District North 24 Parganas, West Bengal) was one of the first persons from his locality to have calluses on his palms and soles. It was around 1984. Arsenic contamination and induced illnesses first reported no more than a year back were yet to be demystified, let alone accepted. At that point of time, manifestations of arsenic-induced illnesses were often confused with leprosy. 'The road used to get deserted whenever I walked down it', reminisces Mr. Islam. His neighbours used to sprinkle Ganges' water all over their houses regularly. The social stigma, undoubtedly, was far more damaging and painful than the calluses and eventually forced him indoors.

---

[1]A total of 132,262 water samples were collected and tested over a period of 2003–2006 from the eight districts of West Bengal under the Joint Plan of Action (JPOA) of UNICEF and PHED. Total number of tube wells in the arsenic affected eight districts are 199,444 which is about 52.5 % of the total government tube wells in West Bengal. Murshidabad district ranks first with respect to the total number of government tube-well installation. Under JPOA total 132,262 government tube wells were tested amongst which there were 76,632 (57.94 %) tube wells having arsenic concentration more than WHO level and 33,755 (25.52 %) tube wells having arsenic concentration of more than 50 μg/L. Government sources show that all the government tube wells were tested for degrees of arsenic concentration status during 2003–07. The density of tube wells is very high in the districts of Murshidabad, North 24-Parganas, Nadia, Burdwan, Hooghly, Malda of West Bengal. Presently these districts are severely arsenic affected.

Several years have passed since then. That terrible disgrace is over. Primarily because during this period the tell-tale skin manifestations of arsenic-induced illnesses have become visible on more and more individuals from his locality. With time, the number of those affected outstripped, by far, those who were not. Obviously the scorn stopped because not many families were left who had no arsenic-affected patient under their roof. Jahirul's son also has the calluses and so does a host of other young men of his age group. Moreover they now know that their illness is due to the arsenic contaminated water they consume. In fact, few lives have also been snuffed out untimely due to the illnesses. But Jahirul does not thank his luck for that.

Jahirul's eldest grandson has recently been diagnosed with pigmentation of the skin. It is just a matter of time before he too joins the ranks of his father and grandfather. And, till date, they do not have a safe water source in their locality. Despite repeatedly voicing pleas and staging agitations and even boycotting polls, Jahirul and his fellow villagers have not been able to bring a safe water source to their village. 'How many more will need to succumb before we will get a water line?' Jahirul's grandson remarked during our visit.

From the Field 5

'I, Dr. K.C. Saha, was the Head of the Department of Dermatology of School of Tropical Medicine, Kolkata (STM) during my first encounter with arsenic-induced illnesses. And looking back, it feels that these happened almost by chance. The first two cases diagnosed were brothers. As a part of my diagnosis process I tested samples from the tube-well they used to drink water from for arsenic. The results came out to be positive. Since this detection of arsenic in groundwater my life sort of changed, overnight. I decided to survey the region from where the first cases were detected. My weekends were free and I decided to utilise those for this purpose. During my visits, I used to carry two cameras and few containers to collect water and dermatological samples.

I started by visiting the village from where the initial two patients had come to STM. To my utter horror, I found many more patients with similar symptoms there. I was informed by the villagers that this phenomenon was not limited to only this village. In fact, at their insistence I visited few nearby villages. And I got a terrible shock. I discovered several more patients in those villages. The fact that they were unaware of their illness was an even bigger problem. It took me great amount of convincing just to get them to visit STM.

I remember vividly one particular case of a victim who was in a particularly bad shape. He had a severe gangrenous wound on his right thumb. I insisted that he must visit STM. He obliged the first time around. Upon pathological testing it was confirmed that the wound was indeed cancerous in nature and that too in a rather advanced stage. I suggested that he must have it operated. However, the patient never returned. When I managed to visit him in few months time his entire right palm was infected. This time around I managed to convince him for surgery. The subsequent amputation halted the progress of cancer for the time being. But I was afraid that the cancer may relapse. And therefore, I advised him strongly for follow ups. However, he never came for those.

I visited his village after a period of 3–4 months. I learnt that he had passed away few days back. I felt sort of responsible for his death. I was aware yet utterly helpless', said Dr. K.C. Saha while elucidating on the initial days of detection of the crisis.

Are We 'Learning By Doing'?

Abstract Ever since the discovery of arsenic contamination of groundwater in West Bengal, there have been numerous interventions. But the sufferings of those affected have not been addressed. To understand this dichotomy better, we need to ask a simple question: What explains the outcome of intervention programmes? To this end we analysed official documents and statistics and interpolated the findings to real-life experiences we gathered during our in-depth field surveys. We tried to comprehend the major lack of cohesion in interdepartmental, interdisciplinary as well as end result related parameters across the entire mitigation endeavour till date.

This is probably the most important question haunting the entire groundwater arsenic mitigation endeavour, since the detection of the crisis, in West Bengal. And unfortunately the answer to this question, in most cases, is a resounding 'No'.

The basic premise of 'learning by doing' is rather simple: are we learning from our successes and failures so we can add on new innovations, and enhance progress by avoiding mistakes? Are we, in this process, finding better solutions to the crisis? Are we identifying the missing links and successfully allocating fresh resources and endeavours towards fulfilling those lacunae?

The government mandate to tackle the arsenic contamination of groundwater problem as a public health engineering crisis ensured emphasis on appropriate mitigation technology development. Global tenders were floated. And technologies did arrive. Primary among those were hand pump attached arsenic treatment units (ATU). Experts have grouped technologies for ATUs into four broad processes (Sharma 2006; Sorg and Logsdon 1978; Mamtaz and Bache 2001; EPA 2000):

1. Precipitative processes
2. Adsorption processes
3. Ion exchange processes
4. Separation (membrane) processes

And for sure, each of these had its own set of advantages and disadvantages. From capital cost to ease of operation to efficiency of filtration and even worse,

A. Das et al., *Socio-Economic Analysis of Arsenic Contamination of Groundwater in West Bengal*, India Studies in Business and Economics,
DOI 10.1007/978-981-10-0682-1_5

generation of toxic waste as a by product—the ATUs had serious issues involving both short and long term viability as well as sustainability!

During our field visits, we found ATUs which have been in operation since installation almost a decade ago. Villagers religiously backwashed (the process of cleansing of the filter to flush out trapped toxins) the filter every day. But, the toxic water generated due to backwash flows directly into the adjacent field thereby contaminating the top soil manifold. Also the chemicals required in the filter, by all means, should have been exhausted long time back. The villagers obviously were oblivious to that fact due to lack of adequate information.

Even today, various funding agencies and government organisations are funding new technologies for field trials. But the fundamental flaw in these technology trials is the fact that once the trial period is over, there is hardly any socio-economic mechanism put in place which ensures these technologies continue serving the people in the optimally efficient manner. There is also a severe flaw in the overall learning curve of technology development and deployment. For example, the ion exchange and membrane processes, two of the relatively more efficient water filtration mechanisms require electricity for functioning. However, electricity supply in many parts of rural West Bengal is till date sporadic at best. In an ideal scenario, technology development should have linked the parameters of optimum filtration mechanism with either intermittent power supply or portable energy sources, e.g. solar power units for sustainable usage. The chemical and spare parts supply mechanism should have also been incorporated in the process by developing a local supply chain.

These major flaws have been replicated in most of the efforts of arsenic mitigation via technological solutions, so far, irrespective of their scale. From ATUs to community level arsenic removal plants (ARP) like those introduced under Swajaldhara water supply scheme reveal similar flaws under scrutiny. To make matters worse, issues of caste, religion and political interference and affiliations have always influenced the decision-making process. In fact, there have been instances where ARPs have been set up in localities with no arsenic contamination as a consequence of the latter three forces. There have also been instances of domestic filters for free distribution lying in stacks in the backyards of government offices while the afflicted population, the intended beneficiaries of these filters, have succumbed to arsenic induced illnesses.

Who Served the Poison?

Abstract What is the source of the arsenic in the groundwater? This fundamental question has been debated by researchers for long. There have been contrasting schools of theories, from geological to man-made, explaining the presence of arsenic in groundwater. In this chapter, through review of literature and interviews with various domain experts, we sum up these diverse theories explaining the cause of arsenic contamination of groundwater.

'Fazal' a common Muslim name literally means the 'bounties from God' Mr. Fazal Karim Mondal of Rajapur village, Rajapur GP, Domkal block, Murshidabad district, West Bengal, however feels that he has received the wrath of God, bountifully. Mr. Mondal is about forty years old. However, everyone assumes his age to be at least twenty years more. How much of this particular phenomenon is due to the reflection of his rather eventful life on his appearance or the symptoms of arsenic induced illnesses are at best a guess?

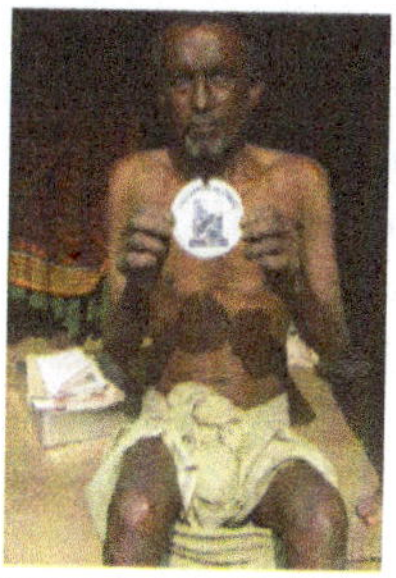

Fazal during our visit in 2010

Fazal had an affluent beginning to life. His house had a personal tube-well, one of the common parameters to judge abundance in rural West Bengal. In due course of time he married Jharna Bibi, a lady whose good looks made Fazal's frame swell in pride. No wonder he caused a few heartburns and maybe collected few curses but

A. Das et al., *Socio-Economic Analysis of Arsenic Contamination of Groundwater in West Bengal*, India Studies in Business and Economics,
DOI 10.1007/978-981-10-0682-1_6

proof of their happy conjugal life can be assumed from their children, two in all. Fazal, at that point in time, was almost living his very own fairy tale.

Sometime soon, however, Fazal ran out of luck. He was diagnosed to be suffering from arsenic induced diseases. And he experienced the beginning of the end, his end. Jharna, rather practically, sought for divorce. Fazal negotiated, or so he thought, the problem and Jharna's demand, by moving out of his ancestral home. They started afresh, in a new location, invested into a deeper tube-well. Prevalent wisdom as well as so-called experts' advice urged him that deeper wells were safe, and he thought the same of his marriage. Initially all seemed to be going well. And then, arsenic found its way up the new deeper well. It brought a reinvigorated threat of divorce from Jharna alongside. Fazal refused, again. But before long, Jharna decided to take matters to her hands. She chose committing suicide to be more appropriate than watching all her beloved die due to arsenic consumption through groundwater.

Fazal however chose to move on. In fact, he married again. This time his bride was Rina, a girl from extremely humble origins. Rina's father was incapable of any physical work and her mother, a labourer in the 'beedi'[1] manufacturing industry. Rina's marriage, for obvious reasons, was a choice of convenience. What's worse? She was kept in dark about Fazal's illnesses induced by arsenic ingestion.

In due course of time, Fazal's condition got worse. He had to sell off all his possessions to foot medical expenses. But the money he got by disposing of his land, bamboo bushes, cows, goats, and pretty much everything else he owned, wasn't enough to keep him healthy. In 2002, a livid and poverty-stricken Fazal had even gone on hunger strike demanding arsenic safe water supply.

Today he is rather quiet and completely invalid. The onus of getting food on plate for Fazal and his four kids has fallen onto Rina. Rina hardly manages to make ends meet by rearing goats. Fazal today rues becoming a father. His inability to earn hurts him like hell. The only time he gets over this feeling of being absolutely worthlessness is when the burning sensation from arsenic induced skin lesions becomes unbearable. Jharna, his first wife, chose poison over consuming water laced with arsenic. Fazal feels he may take the same path sometime soon.

It is now over a decade since Fazal undertook the hunger strike. The act powerful enough to contribute towards India's independence, it seems, has lost its potency. Safe drinking water is yet to reach Fazal's neighbourhood. Nor has any aid, medical or otherwise, trickled in. Fazal's neighbours sum up the situation aptly. They prophesy, "Death is knocking at our doorsteps". In Fazal's, as well as many of his neighbours' homes, hunger strike is no more voluntary, rather obligatory.

What no one is very sure about is that who will help them break the fast and live with dignity and basic human rights?

The first step to break this deadlock is to identify the root cause of the problem. And in West Bengal, the source of arsenic in groundwater is much debated. Essentially, the debate is about identifying the principle culprit on whom square

[1]http://en.wikipedia.org/wiki/Beedi accessed on 10.10.11 at 22:36 h.

blame can be put for the catastrophe. Unfortunately, there has been no clear winner to this argument! On one hand there are people calling for the head of Mother Nature for the disaster, while on the other, an equally vocal lobby has been pointing finger towards human interventions, mostly myopic in nature, like excessive groundwater withdrawal and usage of pesticides and other chemicals in agricultural practices.

In the beginning the blame was rested fairly and squarely on excessive use of insecticides in modern agricultural practices. Resultantly, it was assumed that apart from groundwater, water from dug-wells and ponds adjacent to agricultural fields would also be polluted. But survey results revealed that tube wells were the only source of contamination. Another interesting revelation of the survey results was the fact that quantity of arsenic varies according to depths of the tube wells.

The second breakthrough in the contaminant source identification puzzle came with the confirmation that certain changes in the rock strata trigger the release of the contaminant. And the scientists this time around were looking for geogenic causes as source of the problem. Initially, Rajmahal hills[2] of Chotonagpur plateau was considered to be the primary source for the presence of arsenic in the Ganga–Meghna–Brahmaputra (GMB) plains. A section of geologists are of the opinion that sometime during the Jurassic period, approximately 130–140 million years ago, there was an earthquake beneath the region what is now known as Rajmahal hills. The tremendous lateral strain due to the earthquake forced the arsenopyrite stone strata to a malleable form and essentially spread it out in the basin of Ganga–Padma–Bhagirathi, in a fashion similar to applying jam or butter on your bread! And till 2000, it was the prevalent concept explaining the source of arsenic in ground water. But this spread does not account for arsenic's presence in the basin area of Ganga–Yamuna. And yet the entire GMB plain is severely affected by the crisis.

That's where another school of geologists came to the rescue of the aforesaid theory. They explained that sulphide compounded with arsenopyrite from Gorubathan[3] hills in the Kalimpong subdivision of Darjeeling district, West Bengal, was extended in the catchment area of Ganga–Padma. But this theory is yet to explain the mechanics of the spread of arsenopyrite under the earth's crust.

While theories about spread of arsenic were being debated over across various forums, the contaminant itself chose to appear, in alarming levels, in groundwater of diverse locations. These new appearances simply created a lot of doubt over the validity of the existing theories. The situation in Terai region of Nepal and Bhojpur district of Bihar, few of the latest entrants into the list of areas bearing contaminated sources, cannot be explained with the help of the then existing theories.

To make the situation further complicated, there was arsenic in groundwater in Chandigarh. While plotting the affected regions on a map, it dawned on the surveyors, consisting primarily of a group of scientists from SoES; Calcutta Medical

[2]http://en.wikipedia.org/wiki/Rajmahal_Hills accessed on 28.11.12 at 20:45 h.

[3]http://en.wikipedia.org/wiki/Gorubathan accessed on 28.11.12 at 21:02 h.

College and Hospital; and Institute of Post Graduate Medical Education and Research (IPGMER) under SSKM Hospital that not Rajmahal hill or Gorubathan hills, but the Himalayan mountains are the source of arsenic affecting the entire GMB plain!

On the other hand, some experts were of the opinion that there is no natural cause behind the arsenic pollution in ground water of some places in Murshidabad and North 24-Parganas. The culprit, according to them, was the strainer of the underground pipe of tube wells! But the strainer being made up of brass didn't provide a very plausible argument. Then the blame was directed onto the soldering that binds the pipes together. However, this postulation didn't have too many followers. Not many then, even lesser now.

The other widely prevalent hypothesis for arsenic release and subsequent groundwater contamination is based on excessive withdrawal of groundwater and resultant oxidation of the arsenopyrite. The oxidation hypothesis considers that due to heavy withdrawal of groundwater, aquifer gets aerated. Oxygen thus entering the aquifer oxidises the arsenopyrites rich in arsenic. Acid released as a by-product of the oxidation process leaches the arsenic in soluble form into groundwater.

From the Field 6

The source of arsenic in groundwater is undoubtedly geological. However, the rate at which arsenic is released into the aquifers is influenced by human interference. However, there are two distinct natural safeguarding mechanisms to, at least partially, avert the crisis.

Dr. Tarak Nath Pal, a Senior Geologist at Central Petrological Laboratories, GSI, discovered a safe aquifer in the arsenic prone regions of West Bengal, and named it 'Orange Sand'.

"'Orange Sand', also known as 'Yellow Sand', is a naturally occurring layer of sand found typically at a depth of approximately 120–150 ft below the ground level. It is a rather thin layer and at times is not opted for by the traditional tube-well makers due to lack of flow. However, this layer is probably the easiest short—term solution to the arsenic induced crisis. The water sourced from this aquifer is arsenic free and has been proved to remain so over a period of time", he explained.

His research findings were published extensively. But sadly the intended lessons were not learnt. In areas where there is safe groundwater available for the taking, expensive filtration mechanisms have been put in place to treat unsafe water withdrawn from wrong aquifers. What makes the situation worse is the fact that due to the lack of a well-defined operation and management strategy this expensive filtration infrastructure soon becomes useless junk thereby forcing the exposed population to consume arsenic laced water.

Dr. Pradip Sikdar, Professor, Department of Environment Management, Indian Institute of Social Welfare and Business Management, has been working on the issue of arsenic contamination of groundwater for approximately the past two decades. Dr. Sikdar, during our meeting, pointed out to a very interesting observation.

He asked, "Have you ever looked closely at the cemented platform or of a tube-well spewing arsenic contaminated water?" We had not. But those who will– will find that in arsenic prone regions, nature has devised an easy identifier for distinguishing the unsafe sources. He continued, "generally wells in arsenic prone regions have two distinctly coloured platforms: red and black. This coloration is due to the long-term deposition of minerals drawn out by the groundwater via the hand-pumps. The colour may at times be obscured by the formation of algae etc. but is easily visible by scrubbing off the top muck. The black coloured platforms indicate a significant presence of manganese in the water. If you test the water from such wells, you will find that they usually contain insignificant amount of arsenic. On the other hand, the wells with red platforms indicate danger. These wells produce water that has severe arsenic content and are considerably above the permissible limit. The red coloration of the platforms can be assumed as a sure indicator that such tube-wells must be avoided, at any cost".

However, the manner in which the arsenic contamination of groundwater and the crisis due to this catastrophe have been handled show severe lack of common sense. While there is no doubt that introducing technology intensive solutions have been the preferred mode of mitigation by the government and were also necessary at certain situations. But it surely could have done more to spread awareness regarding simple life-saving findings and observations. Villagers who invested their hard earned money to set up tube-wells to survive the crisis could have been trained to seek out the 'Orange Sand' layer. They could also have been trained to abandon wells with red platforms.

In Love with the 'Devil's Water'

Abstract Arsenic contamination of groundwater in the lower Ganga-Meghna-Brahmaputra plains has been famously recognised as the worst man-made disaster ever. In this chapter, we bring to fore the contribution of various human interventions in fuelling this catastrophe. We look into the circumstances that led to the introduction of groundwater usage. We trace its gradual popularity and also take into account the role of hand pumps and other mechanisms of groundwater withdrawal that may have led to the escalation of the problem of arsenic contamination of groundwater manifold.

Sometime in the 1950s, tube-wells to withdraw groundwater were introduced for the first time in Charamajdia, a village of Nadia district, West Bengal. The villagers with very limited acquaintance with technology fled on witnessing the unforeseen sight of water gushing out through the well screaming 'Devil's Water'. They equated the water coming out from the underground, literally hell or devil's lair, to be a curse. However, subsequently they fell in love with the same due to ease of access. This change of heart was surely attained due to sustained efforts by various interest groups through a series of socio-technical transformations, which can be an interesting research question in itself, but is outside the purview of this book. But one thing was sure, digging of tube-wells to withdraw water for irrigation and drinking purposes became the prevalent socio-technical reality in rural areas.

From the 1950s till the 80s, usage of groundwater was not really in vogue. In fact, groundwater accounted for a mere 40 % of the irrigated area for the period.[1] However this apathy started changing since the 1970s. Green revolution in the form of high yielding seeds empowered by the ubiquitous blessings of groundwater eased the process of conversion. The agricultural productivity with no doubt grew. And no wonder, the shift to groundwater sources started in right earnest and by 2001, there were 5,50,000 tube-wells in use bringing over 47,650 km^2 of land under rice cultivation alone. In fact, tube-well irrigation has completely eclipsed the lift irrigation system that was practised since 1976 in West Bengal. For example, in the year 1976, 20,000 shallow and deep tube-wells were drilled against 700 lift

[1]Refer to Table A.8.

A. Das et al., *Socio-Economic Analysis of Arsenic Contamination of Groundwater in West Bengal*, India Studies in Business and Economics,
DOI 10.1007/978-981-10-0682-1_7

irrigation units from rivers. Subsequently after a period of ten years, the number of lift irrigation units fell to about 69 while tube-well irrigation units went up by an addition of 5634 tube-wells.

The insatiable demand for groundwater has continued to flourish even further. Today, the annual extraction of groundwater in India, in the tune of 210 billion cubic metres, is by far the highest in the world. According to Central Water Commission, groundwater today caters to more than 60 % of the net irrigated area. It accounted for over 85 % of the addition to the irrigated area in the last 30 years.

On the drinking water front, in 1961 only 11,000 villages or 3 % of all villages in India had access to water supply. The rest fulfilled their requirements from surface water sources. The initial push to shift to groundwater in this sector was to avoid the high level of biological contamination found in surface water sources due to poor sanitation and waste disposal and therefore causing tremendous loss of lives and livelihoods. Well, the new water source brought down the instances of waterborne diseases manifold. The villagers, who were reluctant to use it, started referring to it as a status symbol.

Panchayats, the form of local self-government in rural India were given the onus of supplying safe water to its jurisdiction. According to PHED, GoWB, the Panchayats flouted the norms of sinking of tube-wells and set up ten tube-wells using the allocated infrastructure for five. Essentially, it reduced the depth from which water was to be drawn and instead of having one tube-well operating at the depth of 250 feet, two were set up at the operating depth of approximately 150 feet each. The reason for specifying the depth of the tube-well at 250 feet was based on the observation that deeper aquifers were uncontaminated. In fact, there is a clay layer, almost like a natural barrier, which separates the contaminated aquifer from the safe one. However, there have been instances where faulty construction of tube-wells has created pathways for the contaminated water to percolate down to the safe level and in effect polluting an immense resource pool irreversibly. In short, the tube-wells, their faulty constructions and other shortcomings ensured that water was being drawn from an already contaminated source. Therefore, there was water from tube-wells as desired, but the water was not of the ideal quality. The absence of water quality monitoring ensured enough damage was done even before there was any semblance of suspicion.

The official stamp on the hybrid nature of the problem, i.e. caused by the effects of both natural as well as man-made interventions came in 1996. WHO experts Mr. K.S. Subramanyam and Mr. Michel J. Kosnet visited West Bengal for research on the root cause of the arsenic issue. Their verdict, though the problem is geological in nature, has an impact at present ,which is largely induced by human behaviour. Their report indicted the excessive and at times unplanned groundwater withdrawal. Their finding was reinforced by field level survey: during monsoon the arsenic level decreases while in summer it reaches its periodic summit.

The most recent studies have added yet another postulate towards the understanding of the presence of arsenic in groundwater. This theory is based on reductive desorption and dissolution of arsenic adsorbed onto iron oxy-hydroxides in recent sediments, a prevalent condition in the GMB plains. During monsoon, the sediments are subjected to reducing environments with large quantities of rainfall recharge. As redox conditions become increasingly reducing, the ferric iron gets reduced to the ferrous state thereby resulting in mobilisation of some of the adsorbed arsenic. The wetland cultivation technique with increased microbial activity further aggravates the problem.

Therefore, immense groundwater withdrawal during summer months mostly for purposes other than drinking causes lowering of water table and enhances oxidising conditions to facilitate release of arsenic. And during monsoon, the reducing condition enhanced by microbial activities ensures even more arsenic is leached into the groundwater. In a nutshell, arsenic is released all through the year. Its presence, undoubtedly is geogenic, but its release is greatly enhanced by anthropogenic activities.

Theories Elucidating Sources of Arsenic in Drinking Water

- Excessive use of insecticides and pesticides
- Geogenic extraction from sources like Rajmahal and Gorubathan hills
- Himalayan Mountains
- Strainer and soldering of tube-wells
- Excessive withdrawal of groundwater for non drinking water use and resultant oxidation of the arsenopyrite
- Reductive desorption and dissolution of arsenic adsorbed onto iron oxy-hydroxides in recent sediments.

From the Field 7

To feed her burgeoning population was probably the greatest crisis that India faced after independence. The green revolution with mechanised agricultural practices and high yield variety seeds solved the catastrophe, at least on the face of it very conveniently.

However, specifically in the case of West Bengal this independence came at a very high cost. The increase in agricultural produce has been mainly on account of increase in cultivation of Boro[2] paddy in West Bengal. The Boro cultivation is entirely dependent on groundwater; a lot of groundwater. A comparison of water requirements for common varieties of boro rice, wheat and oil seeds may help put things in perspective. The comparative figures are 48 inch per acre for boro cultivation, 12 inch per acre for wheat and 10 inch per acre for oil seeds!

---

[2]http://www.narc.org.np/rice_knowledge_bank/factsheet/boro.pdf accessed on 02.08.2015 at 14:03 h.

This water requirement is fulfilled by shallows run by five horse-power motors, deep tube wells and submersible pumps. The significant lowering of the water table due to agricultural usage is fuelling newer concerns in arsenic affected regions of West Bengal. The vacuum thus created has been proven to expedite the process of leaching of arsenic contained within the bedrocks into the aquifers.

This arsenic-rich water in turn is drawn out for fulfilling agricultural demands and gets deposited on the topsoil. Studies have proved that groundwater containing 100 μg/L of arsenic leaves an annual accumulation of 100 mg/m^2 of arsenic in the topsoil.

There have been recent studies which have proved that same agricultural productivity can be maintained even by reducing water usage. However, the findings of those are yet to become a common practise.

The Assassin on the Move

Abstract Initially consumption of water was considered to be the only source of arsenic ingestion. However, recent studies have pointed out that arsenic intake is not limited to water consumption. In fact, it has infested the entire food chain. Therefore, arsenic consumption is now happening in multiple forms and through diverse avenues. However, the far reaching implications of these additional impact routes are yet to be ascertained. Through literature reviews and in-depth interviews and case studies, we bring forth these additional threats to the already menacing arsenic contamination of groundwater crisis.

Kolkata, the capital of West Bengal, as of Census 2011, has 4.5 million residents. The urban agglomeration, which comprises the city and its suburbs, is home to approximately 14.1 million making it the third-most populous metropolitan area in India. This population, quite naturally, is severely dependent on external produce for fulfilment of its food and nutritional requirements.

In fact, a closer look at the agricultural produce of the arsenic affected districts can be insightful at this juncture. Murshidabad, North 24—Parganas, Nadia, Burdwan, Hooghly, and Malda, the districts mentioned to be arsenic affected in the PHED, GoWB 2006 report, cumulatively produced 58,56,200 tonnes of principle crops in the 2007–08 agricultural year according to West Bengal State Marketing Board.[1] These districts account for over 40 % of the state's agricultural produce. And undoubtedly, a major chunk of their produce travels to feed the hungry millions of Kolkata. Surely, some of the produce finds its way outside the state. After all, West Bengal is the largest producer of rice and vegetables, the second largest producer of potato, fourth largest producer of food grains and seventh in the production of fruits.

What may make you think twice before cherishing the next filling of produce from these fields are the research findings of Dr. D. Chandrasekharam[2] of Department of Earth Sciences, IIT Bombay. He wrote, 'The arsenic content of vegetables and cereals grown using arsenious groundwater (85–108 μg/L) in West

[1]http://www.wbagrimarketingboard.gov.in/index.html accessed on 11.11.12 at 20:00 h.
[2]Chandrasekharam (2007).

A. Das et al., *Socio-Economic Analysis of Arsenic Contamination of Groundwater in West Bengal*, India Studies in Business and Economics,
DOI 10.1007/978-981-10-0682-1_8

Bengal is 300 % greater compared to the mean concentration generally reported in vegetables and cereals elsewhere in the world. For instance, people consuming 100 g of arum, a leafy vegetable, that could contain 0.22 mg/kg of arsenic, will reach the maximum daily allowable limit of arsenic by eating this leafy vegetable alone. Even food cooked with arsenic contaminated groundwater showed high values (0.12–1.45 mg/kg) that fell well above the limit prescribed by WHO.' Few other recent books (Guha Mazumder 2012; Das 2012; and Das and Roy 2015) further document scientific findings on the link between arsenic water and food chain.

However, the safe level of arsenic intake through food is yet to be ascertained by competent authorities. There is no way to prove, beyond doubt, that the arsenic intake via food affects human health in the same way as it does through water. However, in case of food materials there is a definite saving grace. Nature in her curious ways has made sure that the least amount of arsenic finds its way into the food chain. To comprehend this miracle, we need to understand the route[3] taken by arsenic to enter crops. It travels from the contaminated groundwater, via the soil, to the crop irrigated with such water. The accumulation in the specific plant parts is where the miracle manifests itself at its best. Among the plant parts, arsenic concentration decreases as we move from the roots upwards. That means, the arsenic concentration is higher in roots followed by stem and leaves in that order, while the economic or edible parts, in majority of cases, record the lowest concentration of the toxin (Adak and Mandal 2000). Even in the edible parts, for example rice, the majority of arsenic accumulation takes place in the husk leaving the kernel relatively safe for consumption. However, there is more to the miracle. How else can you justify that potato tuber, despite being an underground part (a modified stem), contains relatively lower amount of arsenic (Adak and Mandal 1999)?

The ingress of arsenic into the food chain, unfortunately, does not end here. In Indian farming system, the by-products of agriculture are used extensively for various purposes. For example, the straw and husk produced in rice cultivation are staple diet for cattle rearing. On the one hand, this is a fantastic example of sustainability ingrained in culture. On the other, this is the route through which arsenic enters the food chain of bovines. This in turn, contaminates the milk and milk products which are part of the fundamental dietary supplements in India. The greatest accumulation however takes place in red meat. Red meat, comparatively cheap and voluminous, is one of the primary sources of protein for the poor Muslims in Murshidabad. Its contamination puts the community under even more stress.

However, in Kolkata, the average nutrition being much better, chances of being diagnosed with arsenic-induced diseases get significantly reduced. But who knows how many of them may be sub-clinically affected and may exhibit symptoms of the disease in the years to come!

---

[3]Sanyal and Nasar (2002).

From the Field 8

The primary occupation of the residents of Gharaimari village of Ramnagar in North 24 Parganas is agriculture. From cereals to pulses, fruits to vegetables, they produce it all. And they are quite well-off. Their land is very fertile. They have abundant access to water and other necessary resources, but their greatest asset is their geographical location. They are hardly few hours away from Kolkata, their principal market. In fact, it may not be an exaggeration to say that Gharaimari plays a tremendous role in serving food on the plates of Kolkatans. Kolkata, after all, consumes almost all of the produce from Gharaimari and other nearby villages.

However, Gharaimari is also ill-famous for another reason. It is a hotbed of groundwater arsenic contamination. There have been several instances of arsenic-induced illness and even few deaths. The agricultural fields, located at the fringes of the residential area of the village, are fed sumptuously with the same arsenic laden groundwater which has led to several disasters in the otherwise happy and content lives of the villagers. Till date, there is no permissible limit of arsenic in food in India. And the ill effects of consuming arsenic through the food chain are yet to be ascertained beyond doubt.

Sukanata Ghosh is a farmer from Gharaimari. He has lost his parents due to arsenic-induced illness. He fears for his and family's lives. He is also afraid that the produce from his field may also be contaminated with the same poison. This possesses two fold crises for him. He wonders that he may be responsible for killing silently all those who are consuming his produce. He is also afraid that people may stop buying produce from his fields once they are aware of this contamination. But unfortunately he knows no other way to make a living.

Malignant Intentions

Abstract Arsenic is a proven carcinogen. Prolonged ingestion of arsenic eventually leads to cancer. Chronic arsenic ingestion has rather distinct symptoms. Since the discovery of arsenic contamination of groundwater in West Bengal and subsequent mitigation efforts, emphasis was laid on medical interventions. However, with time, the stress on medical assistance has reduced manifold. The dwindling status of healthcare system has ensured that those affected are even more entrenched in the crisis. Using in-depth interviews of the affected and domain experts and case studies, this chapter explains implications and medical crises faced by the arsenic affected population.

Mr. Sonaullah Mollah's father was a very lucky man. After all, Mr. Mollah, a resident of Srikrishnapur village, Goroimari GP, Domkal block, Murshidabad district was the first individual, way back in 1999, from this locality to travel to Kolkata to receive the luxury of proper medical attention! Well that was all that was lucky about him. His ailments could not be diagnosed in health facilities nearby, let alone be treated. The symptoms however pointed at arsenic-induced diseases. After all, PHED, GoWB in its 2007 report declared all GPs of Domkal to be affected by arsenic contamination of groundwater.

In simple terms, it meant that drinking water from tube-wells in Srikrishnapur is like consuming poison in small regular doses unknowingly. And that is what Mr. Mollah did for approximately 25 years of his life! Inevitably, cancer, in his case, that of liver, caught up with his good luck.

Mr. Sonaullah Mollah tried his best to save his father. He took his unwell father to West Bengal's premiere government hospital, S. S. K. M. Hospital at Kolkata. They undertook the arduous 6 h and 250 odd kilometres long journey on a rickety bus. The road, wherever existed, was bad. The one-way fare was a mere Rs. 82 for the two of them. Or was it not? After all, Mr. Mollah was a small farmer whose yearly income barely added up to Rs. 10,000. This meant he fell marginally below the official Poverty Line Index[1] as devised by the Government of India of Rs. 32.4 a day.

[1]Report Of The Expert Group To Review The Methodology For Measurement Of Poverty, Government Of India Planning Commission June, 2014.

A. Das et al., *Socio-Economic Analysis of Arsenic Contamination of Groundwater in West Bengal*, India Studies in Business and Economics,
DOI 10.1007/978-981-10-0682-1_9

But lack of financial resource or reserve could not belittle Sonaullah's epic struggle to save his father's life. He sold off the last inch of his landed property and got hold of a princely sum of Rs. 1 lac, or equivalent to 10 years of his earning. But that was not going to be enough. They visited SSKM 10 times in the next year or so. Most of the times, they had to return without meeting the doctor. At times, his father was injected with a medicine which had to be bought for Rs. 700. The scheme for free medicine for the arsenic affected did not come to his aid either. He had to procure the medicines from private shops. The princely sum did not suffice the ordeal. His father died and he took all possibilities of a decent survival for Sonaullah and the rest of the family along with him.

Sonaullah today shows symptoms of arsenic-induced diseases, but he is not in a position to undergo medication. Thankfully, he has neighbours sharing comparable predicaments. Their experience with the healthcare system has been analogous and also their reactions are homologous. As if to expedite the eventuality, they continue to drink water from the tube-wells and almost inevitably the symptoms of arsenic-induced illnesses have become visible in quite a few of them.

Consumption of arsenic laced water will impact health severely. It can even kill Mr. Mollah and many other less fortunate individuals like him have found it out over the last three decades of relentless peril. And going by the status of medical research related to arsenic-induced illnesses, it seems, several others will have to find it out with their lives.

At present, there is no effective therapy for this disease. Patients once affected may not recover even after remediation of arsenic contaminated water, with a potential 28.7 million victims in West Bengal alone there are many takers for an effective therapy for chronic arsenic-induced illnesses. However, the fact that the majority of them belong to the resource poor sections of the society reduces, probably, the possibilities of discovery of that elusive effective therapy substantially. The only partial treatment on offer is chelation. But it only helps in excretion of stored arsenic in the body and thereby reduces subsequent cancer risks. And unfortunately, search for effective and economical chelating agents is yet to prove conclusive.

What makes the treatment even more difficult is the fact that with the exception of skin manifestations like melanosis and keratosis, other symptoms and signs of chronic arsenicosis are nonspecific and can occur with other unrelated medical conditions. There is even no direct correlation between arsenic consumption and the onset of typical symptoms. In fact, arsenical skin lesions have been reported to occur in West Bengal after drinking arsenic contaminated water for one year or even less (Garai et al. 1984; Guha Mazumder et al. 1998). While in Taiwan, another hotbed of arsenic contamination of groundwater, the youngest patient drinking arsenic contaminated water who developed hyperpigmentation, a typical skin manifestation of arsenic ingestion, was merely 3 years old (USEPA 1988)! In the Tseng study (Tseng et al. 1968) of an arsenic endemic area in Taiwan, the youngest person with hyperkeratosis was 4 years old while the youngest person diagnosed with skin cancer case was 23 years old! Additionally, the abysmal quality of rural health infrastructure of India raises a valid question: do those suffering have any hope?

From the Field 9

Arsenic-induced illnesses and resultant deaths in any particular locality have a sporadic nature. Not all households are affected. Worse, not all members of the same household are indisposed. In areas where arsenic contamination of groundwater is wrecking havoc, the resultant devastation from a distance seems like that of a game of musical chair. Innocent lives are being snuffed out by arsenic, untimely and randomly.

Mohanpur village is situated next to the ill-fated Srikrishnapur village. Both of them consist of similar number of households having comparable socio-economic backgrounds and religious followings. However, these two villages could not have been fundamentally more different. While arsenic has rendered Srikrishnapur almost into a haunted village, it has not touched a single household in Mohanpur!

It is difficult to find households in Srikrishnapur which have not been aggrieved by arsenic. Many men have died due to consumption of arsenic. Few women have followed suit. On the other hand, in Mohanpur, there was not one instance of this

Table 1 Treatment centres/hospitals for arsenicosis in West Bengal

Teaching	SSKM: STM: CMCH: Burdwan Medical College Hospital
District	Malda: Murshidabad: Nadia: North and South 24 Parganas: Howrah: Hooghly
Sub-divisional	Kalna: Katwa: Jangipur: Kandi: Lalbag: Ranaghat: JNM Kalyani: Barasat: Salt Lake: Barrackpur: Baruipur: Canning: Diamond Harbour: Kakdwip: Uluberia
State general	Baranagar: Bhatpara: Habra: Naihati: Panihati: Sagar Dutta

Source Health on March, 2005–06

Table 2 Arsenic cases and death reports in West Bengal from 1990–2006

Name of the district	Number of Arsenic cases					Number of deaths
	1990–03	2004	2005	2006	1990–06	1990–2006
Malda	1817	150	13	51	2031	5
Murshidabad	2728	23	39	34	2824	4
Nadia	5149	83	38	433	5703	1
North 24 Parganas	1074	26	60	115	1275	10
South 24 Parganas	616	99	38	35	788	17
Howrah	11	0	0	0	11	0
Burdwan	196	13	0	0	209	5
Hooghly	4	0	0	0	4	0
Kolkata	957	2	1	0	960	1
Total	12,552	396	189	668	13,805	43

Source Health on the March, 2005–2006, 2006–2007

catastrophe. The source of drinking water in Mohanpur, like Srikrishnapur, has been groundwater pumped out via hand pumps and tube-wells. In fact, a segment of Mohanpur's population even used to source water from tube-wells in Srikrishnapur!

Today, the neighbouring villages are divided by a chasm of uneasiness. The emotion was summed up by Enuallah Moktar of Mohanpur, "We visit them to extend invitation for ceremonies like marriages and childbirths. While most of the invitations we get are for commemorations and burials. On occasions like Eid and Durga Puja the chasm rears its ugliest head. While one village rejoices and frolics the other reminisces and sheds silent tears." (Tables 1 and 2).

The Opening of the Pandora's Box

Abstract What is the extent of the population affected by the arsenic contamination of groundwater in West Bengal? How is this statistics arrived at? Is the number of affected population constant or has it been rising? How much has the government and its machinery intervened in this public health problem so far? Has the mitigation measures been unbiased in tunes of socio-economic and political parameters? The groundwater arsenic contamination crisis of West Bengal and its mitigation has raised many more such critical questions. In this chapter, adhering to the literature review and evidence-based assessment of secondary as well as primary information, we try to find answers to these all important questions.

In 1983, Dr. Saha's personal crusade to discover the extent of the crisis was lost on the administration. However, his vision and the cause were shared by many like-minded crusaders. Leading the pack were SoES and Dr. Guha Majumder, and their relentless pursuit to uncover the true extent of the crisis led to several uncomfortable facts to surface. For example, in 1983, the number of proven arsenical dermatosis (ASD) patients was 127. Within a span of 5 years, i.e. by 1987, the same number increased to 1217. In other words, there was an almost tenfold growth in the number of ASD. During that period, the greatest number of patients was diagnosed from Murshidabad. Nadia, yet another severely affected district was a close second. In the next 12 years this number grew more than 100 times. North 24-Parganas took the dubious distinction of being the district with the highest number of victims (Saha and Chakraborti 2001) (Table 1).

It can be safely assumed that the endeavour of private players forced not only the state, but also the central government to take into cognisance the plight of the people living in arsenic contaminated groundwater zones in West Bengal. As a result, in February 2002, Mr. Arjun Charan Shetty, the then Central Minister of Human Resource Development, declared the problem of arsenic pollution in West Bengal as a national problem.

A. Das et al., *Socio-Economic Analysis of Arsenic Contamination of Groundwater in West Bengal*, India Studies in Business and Economics,
DOI 10.1007/978-981-10-0682-1_10

Table 1 District-wise report of ASD patients

District	Number of ASD patient	Arsenicosis (estimated cases)
North 24-parganas	1428 (29.35)	60,000 (26.67)
South 24-parganas	472 (9.70)	30,000 (13.33)
Nadia	968 (19.90)	45,850 (20.38)
Murshidabad	1228 (25.24)	59,000 (26.22)
Malda	660 (13.57)	25,000 (11.11)
Burdwan	99 (2.03)	5000 (2.22)
Hooghly*	0	0
Howrah*	0	0
Kolkata	10 (0.21)	150 (0.067)
Total	4865	225,000

Figures in the parentheses represent percentages. *Source* Saha and Chakraborti (2001)
*Not surveyed by the authors of the cited report during this period

Map 1 Arsenic contamination status in West Bengal

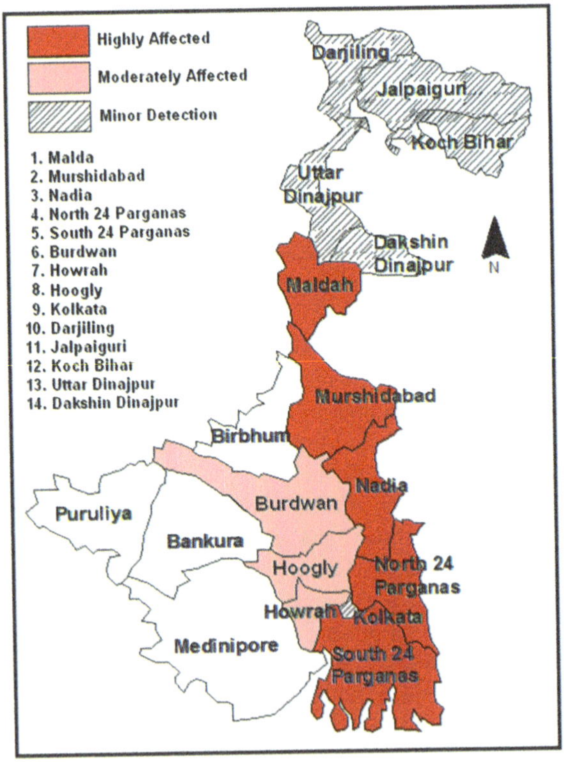

Table 2 Progressive spread of arsenic contamination of groundwater in West Bengal (study conducted by PHED, WB)

Year	No. of affected district	No. of affected blocks	No. of affected village	No. of affected habitations	No. of affected municipalities
May, 1993	7	34	78	149	3
September, 1995	7	56	388	757	9
December, 1997	8	61	1302	3049	9
December, 1998	8	65	1312	3067	9
December, 1999	8	67	1550	3365	11
December, 2001	8	75	2065	4244	11
December, 2002	8	75	2579	4973	11
March, 2006	8	79	3235	6623	11

Source PHED, WB (2006)

Table 3 Progressive spread of arsenic contamination of groundwater in West Bengal (study conducted by SOES, JU)

Year	Number of affected districts	Number of affected blocks	Number of affected villages
1995	6	37	405
1998	8	61	863
June, 2002	9	81	3200
December 2005	12	111	3417

Source SOES (JU) Report (various years)

There have been several efforts to put figures to this misery. One assumption puts the number of people at risk at around 5 million of which 10,134 people (including children) actually being affected (Sengupta 2006). The PHED, GoWB status report 2007 of arsenic in West Bengal (on the basis of 1,32,262 water samples), however puts the total population at risk at a much higher 28.7 million, i.e. 36 % of the total population of 80.21 million (2001 census) of the state. National Institute of Hydrology (NIH) and Central Ground Water Board (CGWB) in their 2010 report puts the affected population at about 16.26 million (20.27 % of the total population of the state) covering 17,533 number of habitations. The anomaly in numbers also extends to that of districts and villages affected. While PHED, GoWB puts the figures for affected habitations at 8 districts, 79 blocks and 3235 villages, the corresponding figures by SoES are 12 districts, 111

Table 4 Number of blocks having arsenic concentration greater than 10 and 50 μg/L

Districts	Area in km^2	Population	Total no. of blocks	Blocks surveyed		As > 10 μg/L		As > 50 μg/L		No. of samples analysed		Maximum As (μg/L)	
				SoES	PHED	SoES	PHED	SoES	PHED	SoES	PHED	SoES	PHED
North 24-Pgns	4094	8,934,286	22	22	19	22	19	21	19	54,368	25,987	2830	3770
South 24-Pgns	9960	6,906,689	29	17	9	12	9	12	9	8333	7284	3700	2720
Murshidabad	5324	5,866,569	26	26	19	25	19	23	19	29,668	40,585	3003	5000
Nadia	3927	4,604,827	17	17	17	17	17	17	17	28,794	29,640	3200	1160
Malda	3733	3,290,468	15	14	7	13	7	9	7	4449	19,279	1904	5000
Howrah	1467	4,273,099	14	12	2	12	2	7	2	1471	879	1333	160
Hooghly	3147	5,041,976	18	17	1	16	1	11	1	2212	2090	600	570
Kolkata	185	4,572,876	–	–		–		–		3626	–	825	NA
Burdwan	7024	6,895,514	31	24	5	12	5	7	5	2634	6518	2230	840
Sub-total	38,861	50,386,304	172	149	79	129	79	107	79	135,555			
Koch Behar	3387	2,479,155	12	5	–	4	–	1	–	474	–	54	NA
Darjeeling	3149	1,609,172	12	4	–	3	–	0	–	562	–	19	NA
Dinajpur (N)	3140	2,441,799	9	7	–	6	–	1	–	990	–	68	NA
Dinajpur (S)	2219	1,503,178	8	6	–	2	–	1	–	452	–	57	NA
Jalpaiguri	6227	3,401,173	13	7	–	4	–	0	–	444	–	27	NA
Sub-total	18,122	11,434,472	54	29	–	19	–	3	–	2922	–		
Bankura	6882	3,192,695	22	17	–	0	–	0	–	279	–	<3	NA
Birbhum	4545	3,015,422	19	11	–	0	–	0	–	718	–	<3	NA
Purulia	6259	2,536,576	20	15	–	0	–	0	–	314	–	<3	NA

(continued)

Table 4 (continued)

Districts	Area in km^2	Population	Total no. of blocks	Blocks surveyed		As > 10 µg/L		As > 50 µg/L		No. of samples analysed		Maximum As (µg/L)	
				SoES	PHED	SoES	PHED	SoES	PHED	SoES	PHED	SoES	PHED
Medinipur (E)	14,081	9,610,788	54	20	–	0	–	0	–	361	–	<3	NA
Sub-total	31,767	18,355,421	115	63	–	0	–	0	–	1672	–	<3	NA
Grand total	88,750	80,176,197	341	241	79	148	79	110	79	140,150	132,262		

Source PHED, SOES (2006)

Table 5 Name of the arsenic-affected blocks/wards in West Bengal

Districts	Blocks/wards as recorded by SOES	Blocks/wards as recorded by PHED
North 24-parganas	Barasat-I, Deganga, Basirhat-II, Baduria, Swarupnagar, Habra-I, Habra-II, Gaighata, Barasat-II, Basirhat-I, Hasnabad, Bangaon, Bagdah, Barrackpur-I, Rajarhat, Haroa, Hingalganj, Amdanga, Barrackpur-II, Minakhan, Sandeshkhali-I	Bangaon, Bagdah, Gaighata, Swarupnagar, Habra-I, Habra-II, Amdanga, Barrackpur-I Barrackpur-II, Barasat-I, Barasat-II, Deganga, Baduria, Basirhat-I, Basirhat-II, Haroa, Hasnabad, Rajarhat
South 24-parganas	Baruipur, Sonarpur, Jaynagar-I, Mograhat-II, Bhangor-I, Bhangor-II, Budgebudge-II, Bishnupur-I, Bishnupur-II, Canning-I, Canning-II, Mograhat	Budgebudge-II, Bishnupur-I, Bishnupur-II, Baruipur, Sonarpur, Bhangor-I, Bhangor-II, Jaynagar-I, Mograhat-II
Murshidabad	Baharampur, Bhagabangola-I, Bhagabangola-II, Beldanga-I, Beldanga-II, Suti-I, Suti-II, Raninagar-I, Raninagar-II, Domkal, Jalangi, Hariharpara, Nawda, Raghunathganj-I, Raghunathganj-II, Farakka, Murshidabad (Jiaganj), Lalbag, Lalgola, Samsherganj, Kandi, Sagardighi, Bharatpur-I, Khargram	Farakka, Samsherganj, Suti-I, Suti-II, Raghunathganj-I, Raghunathganj-II, Lalgola Bhagabangola-I, Bhagabangola-II, Raninagar-I, Raninagar-II, Jalangi, Domkal, Murshidabad (Jiaganj), Baharampur, Hariharpara, Nawda, Beldanga-I, Beldanga-II
Nadia	Chakda, Santipur, Nabadwip, Tehatta-I, Tehatta-II, Haringhata, Kaliganj, Karimpur-I, Karimpur-II, Hanskhali, Ranaghat-II, Nakashipara, Chapra, Krishnanagar-I, Krishnanagar-II, Ranaghat-I, Krishnaganj	Karimpur-I, Karimpur-II, Tehatta-I, Tehatta-II, Kaliganj, Nakashipara, Chapra, Krishnanagar-I, Krishnanagar-II, Ranaghat-I, Krishnaganj, Chakda, Santipur, Nabadwip, Haringhata, Hanskhali
Malda	Kaliachak-I, Kaliachak-II, Kaliachak-III, Englishbazar, Manikchak, Malda (Old), Chanchal-II, Ratua-I, Ratua-II	Kaliachak-I, Kaliachak-II, Kaliachak-III, Englishbazar, Manikchak, Ratua-I, Ratua-II
Howrah	Bally, Jagacha, Uluberia-I, Uluberia-II, Sankarail, Amat-II, Bagnan-I, Bagnan-II, Amta-I.	Uluberia-II, Shyampur-II
Hooghly	Balagarh, Canditala-II, Dhaniakhali, Goghat-I, Haripal, Khanakul-I, Khanakul-II, Pandua, Polba-Dadpur, Singur, Sreerampur	Balagarh
Kolkata	Ward number: 1, 32, 55, 67, 71, 82, 83, 88, 89, 92, 93, 95, 96, 97, 98, 102, 103, 104, 110, 111, 112, 113, 116, 117, 118, 120, 121, 123, 139, 140, 141 (31 wards having AS > 50 μg/L out of total 141 wards)	Not reported

(continued)

Table 5 (continued)

Districts	Blocks/wards as recorded by SOES	Blocks/wards as recorded by PHED
Burdwan	Purbasthali-I, Purbasthali-II, Bardwan-I, Kalna-I, Katwa-I, Memary-II, Raina-II	Purbasthali-I, Purbasthali-II, Katwa-I, Katwa-II Kalna-II
Kochbehar	Kochbihar-I	
South Dinajpur	Balurghat	
North Dinajpur	Goyal Pokhar-I	

Source Personal communication with PHED, SOES

blocks and 3417 villages. The rational explanation for this variation in reports is the different number of samples tested by these different bodies.[1] But, thirty years after the discovery of the crisis not having a definite number for polluted sources and population at risk probably makes a statement about the way the crisis has been tackled so far. (Map 1 and Tables 2, 3, 4, 5).

[1]Refer to Tables A.7a and A.7b.

Spilling the Beans

Abstract Since, the detection of the groundwater arsenic contamination crisis mass media have played a critical role in spreading awareness about the catastrophe. Due to their sustained reporting and efforts, awareness level has increased substantially. But, sustained communication, especially by the state and its agencies, has not happened. In fact, the state's response has been largely reactive due to its unwillingness to accept and therefore address the crisis by reviewing literature and media reports over this period, we scrutinise the role played by mass media in spreading awareness and thereby contributing towards mitigation of the arsenic contamination of groundwater crisis.

Dr. Saha's reports to the state government on arsenic contamination of groundwater, initially, met with reactions ranging from utter disbelief to disdain. However, undaunted by the reactions, he carried on. SoES, Dr. Guha Majumder and several others joined the cause. Eventually, the print media started reporting about the plight. The crisis, no longer could be swept under the carpet. It was this consistent barrage of academic and media reports which finally forced the government to move out of its slumber and accept that there was indeed a grave problem.

The role played by media in spreading awareness regarding arsenic has been commendable. Their relentless reportage of the catastrophe ensured that the issue could not slip into oblivion like many other public health issues. The largest number of stories dealt with scientific enquiry regarding the source of arsenic and different filtration technologies. A trend which reflected the way the crisis was perceived and was tackled by the concerned authorities. However, under scrutiny the newspaper reports bring forth few interesting insights. Issues addressed ranged from survey reports by different mitigating as well as research-focused agencies on the presence of arsenic in groundwater; to number of persons affected and afflicted by arsenic-induced diseases; to number of deaths due to arsenic-induced cancer. There were few reports regarding the social and economic loss due to the catastrophe and also about different court cases.

The sustained reportage encouraged many patients and their families to come out in the open. But, somehow it failed to diminish the sense of stigma attached to the

A. Das et al., *Socio-Economic Analysis of Arsenic Contamination of Groundwater in West Bengal*, India Studies in Business and Economics,
DOI 10.1007/978-981-10-0682-1_11

issue. Reports point towards several socioeconomic tragic consequences of the crisis. For the first time, it emerged that the cost of the catastrophe was no longer limited only to health effects. Socioeconomic loss due to the disaster was far intense. There were instances where wives were deserted by in-laws within weeks of marriage due to dermatological calluses; agricultural workers were not being allowed onto the fields; and individuals being ostracised by community because of skin manifestations. It may not be an exaggeration to state that the victims of arsenic-induced illness suffered equally, if not more, from the appalling socioeconomic situation. In some way, arsenic-affected locations became concentration camps where the affected families were held as hostage in their own homes. The discovery of newer-contaminated sources and the rise in social stigma towards those who are affected kept pace with each other.

To curb the disturbing trend of social stigma and to build capacity amongst all stakeholders the PHED set up Communication and Capacity Development Unit (CCDU) on 10th November, 2004. One of the primary responsibilities of CCDU was to initiate information, education and communication (IEC) activities towards mitigation of the crisis. The long-term goal of the unit was to help develop community ownership of water sources and the various filtration systems those were installed to make safe water available. The target was also to help develop skills related to monitoring and surveillance those were intrinsic towards achieving this target.

The communication programme aimed at addressing two distinct types of audiences. Implementing agency members, NGOs and members of Panchayat constituted the first group. The second comprised of villagers in general and women in particular. However, effectiveness of this initiative left much to be desired. In fact, during our field visits we rarely came across village water and sanitation committees as proposed through the communication initiative.

Moreover by interpreting data from our field studies in the affected areas, we realised that the source of information about arsenic for most of those surveyed has not been programmed by CCDU (Table 1).

Table 1 Sources of information about arsenic in survey areas

Sources of information	Number of households/ respondents	Percentage
News paper	6	3
TV	2	1
Radio	4	2
Residents of the village	66	33
Relatives	31	15.5
Government	14	7
NGO	5	2.5
Others (researchers, surveyors from SoES, BESUS)	72	36

Further confirmation of the inadequacy of the awareness generation programmes can be derived from the fact that only 14 % of the respondents remembered events of the same nature in their locality over the past 6–7 years. They were sure that no such event occurred in the past year. In fact, information regarding the catastrophe was attributed to word of mouth and visits by academics. No wonder, the demand by those affected for safe water, even after three decades of the detection of the crisis, has not gathered momentum.

From the Field 10

Mr. Debdut Ghosh Thakur is a respectable journalist of a leading Bengali daily. He is a fiercely honest, old school, reporter who has played a significant role in spreading awareness regarding arsenic contamination and induced health hazards. Mr. Ghosh Thakur has travelled extensively through the arsenic-affected regions to report about the crisis. He has seen many a tragedy unfold. However, he never probably would have imagined himself being at the centre of one.

His reportage about government not complying with the court's order to uproot arsenic spewing tube wells forced the state government to move into action. The government uprooted one such tube well in Binimay Para. Mr. Ghosh Thakur went to visit the locality some days later. Mentally, he was prepared for great adulations. However, at Binimay Para a rude shock awaited him. He was held hostage by the community for over 24 h. They were seething in anger about the reportage and subsequent actions. The government did take away the contaminated source and saved the villagers from consuming arsenic-laced water. But, it failed to provide for an alternative source. Resultantly, the villagers were left with no source of drinking water and had to resort to sourcing water from surface water pools which teemed with biological and bacterial contaminants!

Threats to Sustainability

Abstract The groundwater arsenic contamination is over three decades ago. Over the years, there have been slew of mitigation measures planned and executed by several private organisations as well as government bodies. Unfortunately, none of these have attained the elusive mark of sustainability. To decipher the crucial missing link using theoretical framework and observations from the field, we make an assessment of mitigation actions taken over the past three decades. By scrutinising these efforts, we try and comprehend the real reasons due to which the mitigation efforts have failed to yield the optimum results. In the process, we also try and reinforce the fact that the arsenic crisis should always have been considered and addressed as a socio-economic crisis rather than only a public health engineering one.

The major problem with all arsenic remediation measures so far has been their inability to provide long term access to arsenic safe water. The critical question that arises therefore is that what is missing in making the solutions sustainable. Looking back, one can safely conclude that the entire exercise was flawed due to one basic error in the business or diffusion model at least during the first phase of mitigation exercise. The entire mitigation effort was conceived as an expenditure model without making sure the sustained flow of capital requisite for maintenance of the introduced technology and subsequent distribution of arsenic safe water (Das 2012). In a nutshell, the challenge was limited to finding technology-based solutions to a perceived public health crisis.

However in the latter phase of mitigation efforts, there were few positive changes. For example, public private partnerships were introduced to facilitate the implementation strategy. Under these partnerships, the government still paid the capital cost to private partners to instal the technology. But the terms of engagement were reworked to introduce components wherein the technology providers had to mandatorily sign up contracts for three years of maintenance in collaboration with the NGOs. The private partners and NGOs were given the mandate to establish a community-based institutional arrangement to ensure day-to-day operation and maintenance.

A. Das et al., *Socio-Economic Analysis of Arsenic Contamination of Groundwater in West Bengal*, India Studies in Business and Economics,
DOI 10.1007/978-981-10-0682-1_12

With this new agreement, private partners did come in as their risk of return was covered by government paying for their capital investment. The private investors got on-the-spot support as site selection and all other installation-related activities happened with active PHED, local government and community involvement. In the light of these observations, a simple question arises: What did then really go wrong in the mitigation effort or is everything right and solicits replication elsewhere?

Based on our extensive field level experience, it may not be an exaggeration to state that the mitigation exercise was conceived and actualised in a socio-economic-institutional vacuum. Technology installation happened but its social embedding was not given its due importance. While importance was rightly ascribed to development and identification of appropriate technology and resultant safe water supply. But somehow, the social dynamics of cooperation that make the introduced technology work and the actual cost of the disaster borne by the society were, probably, lost upon the authorities.

It may be worthwhile at this point of time to comprehend the meaning of sustainability in the context of arsenic contamination of groundwater. This will also help us better understand the shortfalls of the arsenic mitigation efforts. This will also eventually help define a mitigation plan that can strive for sustainability in the future efforts.

Sustainability can be achieved by putting in place, an appropriate combination of natural, human, man-made and social capital. In the arsenic contaminated regions of West Bengal, the natural capital, i.e. groundwater stock is available, but its quality is grossly unacceptable from heath standards. On the other hand, by extending the ambit of natural capital to include water stock from all sources, we arrive at an interesting observation. Gangetic West Bengal enjoys an average annual rainfall of approximately 1439 mm. The Ganges itself carries an impressive water load. However, only a part of this is utilised for conservation, distribution and subsequent consumption. The flow of Ganges also involves the question of downstream allocation cutting across borders. On the other hand, the overt emphasis on using groundwater has, at least partially, aggravated the natural process of arsenic leaching into groundwater in the absence of any institutional framework and mechanism of regulated extraction.

The most basic parameter for defining the standard of human capital in a locality, in this particular context, will be the status of health of the occupying population and the technical knowledge imbibed by them to run the technologically advanced arsenic water treatment systems. During our field survey, we came across several families which are suffering from arsenic induced illnesses. In some families, the impact of arsenic induced illnesses was found to be worse than others. Subsequently, there were families where all earning members were either incapacitated or dead leaving the surviving members in dire economic health. Not to mention, they were also the victims of the contamination and therefore had reduced working capacity of a meagre two hours per day (Roy 2008). In fact, interpreting data from our field surveys, it may be concluded that there have been instances where income generation potential of the affected has reduced by half over a short period of three years. What makes this drop even more alarming is the fact that these families were bordering along the poverty index even prior to being diagnosed with arsenicosis. The drastic reduction of income simply meant that they went from bad to worse, if not plummeted to the worst case scenario. For these families, low labour potential due to arsenic induced illnesses meant lower income generation which in turn accounted for an even lower level of nutrition. Simply put, the labour potential was steadily declining for even the subsequent generations aggravating the crisis and reducing the possibilities of breaking free from this downfall. Essentially, these families were trapped into a vicious cycle of poverty triggered off by one of the primary requirements for life: drinking water. In their case, the water considered a right and otherwise a free resource dragged their sorry state of affairs to abysmal levels (Fig. 1).

It may be worthwhile to extend further the relationship between the natural and the human capital at this juncture. The share of residential sector is a meagre 7 % of India's total water requirement (NCIWRD, Report, 2010). However, the impact water wields on health, especially in arsenic affected regions demands nothing less than special emphasis on the mutual relationship between natural and human capital. It is also important to keep in mind that in context of sustainable development, the relationship between safe water and human progress has been proven to be inseparable. And the loss of human welfare due to drinking of arsenic contaminated water has been estimated to be on average Rs. 300/- per month per family (Roy 2008).

As of now, the rural areas lack the required human capital for managing the arsenic safe water treatment systems. The scenario becomes even dire, if we include the safe disposal of sludge generated due to treatment of arsenic contaminated water with no long term effect into the equation.

The principal goal of sustainable development is to achieve equity. The factors resulting in inequitable distribution of water are manifold. Principal among those

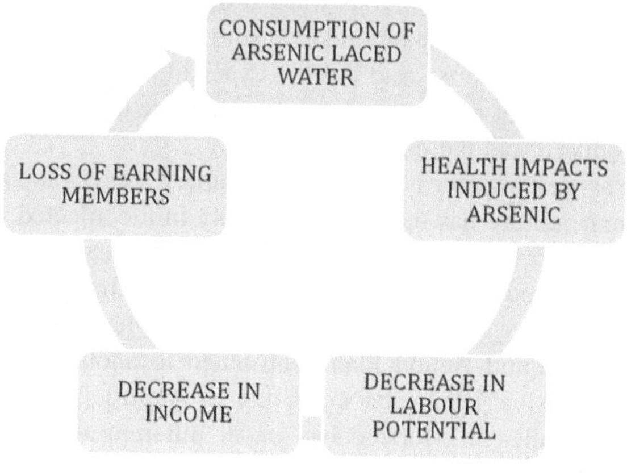

Fig. 1 Vicious cycle of poverty induced by consumption of arsenic laced water

are institutional shortcomings, market failures, informality of markets, price distortions and wrong price points. It is imperative that these factors be comprehended in totality in arsenic prone regions before drawing up the future course of action. These factors also incidentally create the framework of man-made capital.

In arsenic prone regions, the major impetus in mitigation efforts has so far been dedicated towards field testing of various arsenic removal devices. Due to a lack of societal sensitivity of implementation programmes in the absence of complimentary communication and resultant demand generation, the technologies have always remained alien for the community. In other words, these technologies did not get integrated into the collective knowledge of the community. Therefore during field survey, we came across several relics of these now-defunct filtration machineries lying abandoned.

However, there have been quite a few innovative local usages of these structures that fit into the social needs. For example, parts of a filter kit have become grain storage containers. The pipelines have become posts to tie cattle for grazing. Whereas, on papers, contractually empowered committees were formed to scrutinise the on-field technological efficiency of the filters. But, there were no such efforts by the committees to ensure end-user readiness check for managing an equitable distribution of safe water in a given community. Therefore, access to filtered safe water was often held hostage to a number of apparently irrelevant but locally relevant socio-economic factors such as political alignment, religious following and caste system.

The mitigation process in its fragmented approach also ascribed somewhat limited emphasis towards finding medical solutions or at least provisions for interim relief and diagnosis to already ailing patients. However, there were no efforts to comprehend the socio-economic cost of the disaster, which, by the way was the biggest hurdle towards equitable safe water supply. Therefore, the ill-conceived and planned mitigation measures essentially achieved to incur expenditure at both ends of the spectrum. While the administration spent from its own coiffeurs as well as borrowed billions to develop ambitious water testing and supply infrastructure, the affected community spent in cash or kind to ensure mere survival on alternatives not connected to mitigation efforts. And despite expenditure incurred by both parties, the resultant return, to put nicely, has been minimal due to disconnect between the technology introduced and the end-users' society.

Our surveys as well as information from secondary sources indicate clearly that duplication of expenditure has occurred repeatedly in the affected regions.

The first lot of 265 tube-wells, post the discovery of arsenic contamination of groundwater, in Murshidabad was carried out under the First Phase of Action Plan during 1994–95. Another 205 were sunk in the Second Phase of Action Plan during 1997–98. The State Fund Action Plan contributed to another ten, while PMGY accounted for a massive 1952 tube-wells. Up to August 2003, total 2329 deep tube-wells were installed in Murshidabad under different arsenic mitigation programmes. This number grew to 2432 in April 2007. Effectively in a dozen years, almost 2500 tube-wells were sunk in Murshidabad alone. Another 114 dug-wells were constructed up to 2007. Apart from these, a total of 1125 ATUs were set up and 47 PWSS were sanctioned till that point in time. However under scrutiny, sanctioning of few of the projects defied logic. For example, no ATU was installed in Raghunathganj-I block despite the fact that it had one arsenic affected mouza as per PHED report (Das 2012). However, Sagardighi block got six of these filters and had no instances of arsenic contamination as per the same report.

The overall picture for West Bengal shows an eerie similarity to that of Murshidabad. In all arsenic affected districts, interventions in the form of technology driven filtrate mechanisms of varied scales were established.[1] However in majority of cases, the mistakes those were prevalent in Murshidabad were rather apparent.

For the purpose of our survey, during the year 2007–2009, we visited 70 ATUs all over the district.[2] To our horror, we found that only five of them were functioning at the time of visit in the period of May 2009 to July 2010. There have been few cosmetic efforts to involve the community in the operation and maintenance of the ATUs mainly because of the conditions laid out by funding agencies. However, in most cases, such casual attempts to form community-based committees could not integrate or transcend the social division and hindrances of cast, community, religion, political affiliation and at times even gender. Technically, ATUs needed chemical replacement after a certain period of time. The community, who was to take this responsibility in most cases, could not come to consensus regarding who will bear the cost for the same. Also, they lacked adequate skills to maintain the systems. It was ignored at the project planning stage that how mandatory steps like back washing and sludge disposal will be done by informal community-based committees who had no such exposure so far. Rather obviously, community cooperation-based institutional arrangements to maintain the advanced technological systems failed, in most cases, in an otherwise fragmented community. However, members of the communities surveyed by us still had to travel for 27 min daily, on an average, to collect safe water. Moreover, many of them have availed the services of rickshaw pullers to get their daily fill of guaranteed safe water supply for a price A mid-term appraisal[3] of the fifth Five-Year Plan mentioned 52 min to be the average time spent by Indians to collect water from safe source. In river infested and rain fed West Bengal, the corresponding figure to be in a similar range shows failure of strategy and governance for rural water supply system (Fig. 2).

Evaluating secondary data to corroborate our field survey results, we found an exact albeit unfortunate match. For lion's share of mitigation efforts carried out by different stakeholders, evaluation process while considered costs incurred and inputs engaged in production process, but grossly overlooked the demand side of

[1]To know more refer to Table A.11.

[2]Detailed field experience of performance of mitigation technologies is available in Das and Roy (2013).

[3]Evaluation Study On Rajiv Gandhi National Drinking Water Mission (RGNDWM) by Programme Evaluation Organisation, Planning Commission, Government of India, New Delhi, November, 2010.

Fig. 2 Photographs of some defunct ATUs lying here and there in the survey area. *Source* Das (2012)

7. Near the house of Rejaul Haque (Village:Rakhaldaspur)
8. Near the house of Haji Khadija Bewa (Village: Shyamdasdiar)
9. Near the house of Nijarul Islam (Village:Amirabad)
10. Near Shyamdasdiar Primary School (Village:Shyamdasdiar, Shialmari)
11. Near Kadamtala Primary School (Village:Kadamtala)
12. Near the house of Abdul Bari (Village:Amirabad)

Fig. 2 (continued)

the story that reflects the return that can be generated from the said expenditure[4] for reinvestment in physical technology, its maintenance and eventual replacement.

[4]See Technical Appendix: Threats to Sustainability.

From the Field 11

Sushanto Nashkar is a brick kiln worker from Rajapur village, Domkal block of Murshidabad. With over 10 years of experience under his belt, he is a veteran skilled professional of his trade. However, his income has been steadily decreasing over the past few years while the wages, for a change, have increased. Mr. Nashkar is a victim of arsenic induced illnesses. He has painful calluses on his palms and suffers from shortness of breath. He also cannot stand the scorching summer sun, which unfortunately represents the peak season for brick production.

Mr. Nashkar, at his most efficient, was earning on an average Rs. 3000 monthly. The income was just enough to put food on the plates of his six-member family. Presently, his income has dwindled to under Rs. 1500 monthly. His eldest daughter is of all but 12. But she has to join work to supplement her father's income. Dropping out of school was part of her sacrifice. However, their combined income, at times, is insufficient to feed the family.

What has hurt them more is the expenditure generated on the account of medicines for Mr. Nashkar. He uses ointments to soothe the inflammatory sensation of his palms. The temporary relief it provides allows him to work for a few more hours and resultantly earn a bit more. But, ironically, the extra money he thus gets is oftentimes less than the expenditure incurred to earn it. Mr. Nashkar today seriously is in a dilemma. He wishes for his death but is confused whether his death will revive his family's dwindling fortune. This is a live example of how depleting human capital can make income growth unsustainable.

From the Field 12

India is a democracy. In other words, the power lies with the people to live contently the way they want. They retain the right to chose what is right for them. But for Mr. Rafiqul Islam of Khidirpara village, Raipur GP, Domkal block, Murshidabad it is not so. The reason for his misfortune is his political belief. Or thats what his experience has given him to believe.

Mr. Islam's village has arsenic safe water sources. But unfortunately, the stand posts have been installed in the houses of the people with affiliation to a certain political party. The safe water, though public property and a fundamental right, is not publicly available. Rafiqul has tried to get water from these sources many times but had no luck. He had to come back empty-handed even after hours of waiting in the queue.

'Without the right political party affiliation you do not have access to the right quality of water', he sums up in a dejected tone. He has stopped trying since. Many of his neighbours share the same predicament. They drink water from contaminated tube-wells. 'At least you save yourself the humility', he continues, 'we have already chosen death over humility.' And the idea of democracy takes a severe beating with this choice.

The Value of Human Life

Abstract What is the human welfare loss due to arsenic contamination of groundwater? Can this loss be quantified? How much is an individual as well as a family from the affected region willing to pay to save themselves? To understand the real price paid by those affected due to the arsenic contamination of groundwater we adhered to analysis of data collected during field surveys using econometric model. Through this endeavour, we intended to ascertain that the burden of the disease is much more than health effects only and has severe socioeconomic implications.

What is the actual impact of the disaster if the currency of evaluation is the sufferings of those affected? Are damages to health leading up to death the only consequence of the disaster? How do we evaluate the actual sufferings of the affected population: individually and socially, monetised and non monetised, over a lifetime if not dying? For starters, if we consider the narrow view of humans as 'economic beings' then we need to assign numbers, unfailingly, to the monetary cost borne by the sufferers for continuing to live with a certain dignity. In other words, we need to define the willingness to pay (WTP) for arsenic safe water for those already affected by the disaster to avert the same. Those who are already having health damages are also spending in order to seek cure or to save on loosing precious productive time. These damages also definitely represent these individuals' as well as these families' WTP. In order to quantify this cost, we need to resort to alternative methods of economic valuation. Water used by most of the rural households is not allocated through a regular market institution where price mechanism is directly observable. So the challenge is to get to know the value attached by users to water in the absence of an explicit market. But, we know that there is enough information available which individual consumers, even in this rural society, can and do process and hence are capable of differentiating water qualities from alternative sources. Conventional definition of WTP is the maximum amount a person would be willing to pay, sacrifice or exchange in order to receive a good or to avoid something undesired. The undesirable incident which the affected are willing to avert, in this particular case, is consumption of arsenic-contaminated water, either directly or indirectly. We also

A. Das et al., *Socio-Economic Analysis of Arsenic Contamination of Groundwater in West Bengal*, India Studies in Business and Economics,
DOI 10.1007/978-981-10-0682-1_13

extend the definition and try to comprehend that the amount the affected is willing to pay to secure a defined improvement in his conditions. This will give an idea about the economic valuation of the improved service; in this case it is provision of safe water from the perspective of those affected. This crucial piece of information will also be the premise upon which future policy decisions need to be based upon. It will also provide a blueprint to calibrate all future allocations of public funds. It will ensure economic incentive is ingrained into the rubric of all future mitigation measures which in turn will yield sustainability.

To determine WTP[1] we conducted an in-depth survey involving 200 households[2] at Jalangi and Raninagar-II blocks of Murshidabad. These, according to PHED, are the worst affected blocks in terms of arsenic contamination of groundwater.

The theory assumes that individuals have well-defined preferences among alternative bundles of products comprising both market and non-market goods of varying quantities. It is assumed that the consumers know their preferences and they consider substitutability among alternative goods in a bundle. Substitutability implies that if the individual reduces the consumption of one good, then it is possible to increase the quantity of another good in the bundle so as to leave the individual equally satisfied.

During the survey[3], respondents were asked whether they were willing to get arsenic safe water for their family and pay for the operation and maintenance (OandM) costs on account of those mitigation measures. All the respondents agreed to this proposition of securing arsenic safe water.

Subsequently, they were asked to put a number, i.e. price, they were willing to pay to avail better services. They were asked to choose a figure between Rs. 10–100 per month for the same. The survey results show, expectedly, that as contribution charge increases, number of contributors decreases. About 8.5 % of the respondents cited their incapability to pay any amount. The average WTP towards O&M of mitigation measures, according to our survey, was Rs. 26 per month. In relation to income and expenditure, low-income (BPL) households were willing to pay substantially greater proportion of their incomes for availing better services. BPL households were willing to pay 1.1 % of their income for the services, as against 0.99 % in case of middle income group, and 0.27 % in case of higher income group. In terms of expenditures also BPL households were willing to pay more than their counterparts.

The apparent contradictory finding has an interesting message to deliver. This study is a fair indication that the affected population belonging to the BPL category is already incurring higher indirect costs, at least in our survey area, and in turn, is willing to pay for better services. One must remember, the WTP figure thus arrived

---

[1]An earlier attempt for North 24 Parganas district of West Bengal is in Roy 2008.

[2]Please see Technical appendix to this chapter for technical details of estimation of WTP.

[3]Read more about the Willingness to Pay survey at Technical Appendix: Methodology.

does not take into account the indirect and hidden expenses already borne by the respondents. In fact, the effective WTP can be arrived at only after summing up the direct cost of O&M for safe water supply and the various indirect costs, e.g. travel cost for safe water sourcing; adapting cost on account of medical expenditures; averting cost on account of domestic filtration mechanisms and hidden cost on account of social cost of exclusivity. The indirect costs are already borne by the affected population, and at times, their cumulative impact is much more than the direct burden of the disaster. In fact, it may not be an exaggeration to state that the indirect costs of averting the crisis have been the principal burden on the affected population. Women, the traditional caregivers in Indian society have borne its brunt.

Globally it is estimated that women and girls in low-income countries spend 40 billion hours every year on account of fetching and carrying safe water from far-away sources (Blagbrough 2001). It has been proved that when women and girls have easy access to a clean water supply, girl children lose less school time (WSSCC 2004). But to seek safe water in Murshidabad, where 64.53 % of the total area is reeling under arsenic contamination, is not a cakewalk. The expenses incurred to source safe water can either be direct, i.e. paying people for their service or indirect, i.e. spending time to fetch it. For example, the research population covered an average distance of 59 m daily to fetch water for consumption. The average time spent for the same purpose was 27 min. Putting into perspective the fact that all-weather roads are yet to become a reality in most parts of the arsenic-infested regions in Murshidabad, this is no mean feat. In fact, nagging injuries due to carrying heavy loads were also reported by several women of the survey population.

The cost incurred on account of seeking treatment is yet another burden on the already resource poor community. The act of availing medical attention in most of the arsenic-infested regions begin by traversing the distance between their residence and the health facility. In most cases, this distance is quite substantial. This journey often results in incurring auxiliary expenditures. The primary repercussion of this expense, is not to avail it, especially for the women who are usually not the head of their families.healthcare. Even for the men who travel some 250 odd kilometres to Kolkata to avail tertiary care, the expenditure was no less daunting. In our study all people who sought treatment at Kolkata were males and belonged to the upper or middle income group. The only thing the poor could afford was to die.

Apart from affordability, the visit to the city for most of these people meant taking a great leap of faith. Most of them have never been to Kolkata. Therefore, they depended on relatives and tout who have visited the city before. In other words, the patients had to bear for additional expenditure on account of travel, and food and boarding costs to accommodate these co-travellers. Moreover, they were exposed to hardships like spending nights on the footpath. Overall, the entire process was and still not very conducive for those seeking treatment. Moreover to foot these bills, the families had to cut down expenditures on account of other essential requirements. In the study area, we found families had to borrow money from relatives and even sell assets to foot the medical expenses.

Table 1 Management of money for treatment-related expenditures

Ways	Number of households/respondents reported money management		
	For male	For female	Total
Reducing expenditures	39 (92.86 %)	3 (7.14 %)	42 (54.53 %)
Selling assets	2 (66.67 %)	1 (33.33 %)	3 (4.11 %)
Taking loan	0	0	0
More than one way	25 (89.29 %)	3 (10.71 %)	28 (58.36 %)

Take for example, the case of Mr. Alauddin Mondal from Khidir Para village, Domkal block. Since the onset of the skin lesions some 16 years back he has visited Kolkata some 11 times. Each time he was accompanied by at least one person. The expenses incurred due to this co-passenger for each trip has been a minimum of Rs. 250. Today, Mr. Mondal considers himself unlucky because he is still alive. He says, 'At least, my family would have some assets left had I died 16 years earlier. Now, they are like paupers and I will certainly die, eventually.' (Table 1)

Pricing the Life?

Abstract To arrive at the optimal price that the affected are willing to pay in arsenic affected regions of West Bengal in order to continue in leading a healthy life one needs to identify various hidden costs. To identify these costs we undertook in-depth socio economic surveys in arsenic infested regions of Murshidabad, West Bengal. Findings from this survey not only helped to determine the method to arrive at the most viable value for willingness to pay, it also revealed the missing defence mechanisms, absence of which, were escalating this cost and eventually forcing those affected into a vicious cycle of poverty.

A variety of techniques and models have been developed for assigning monetary values to gains or losses in human satisfaction associated with changes in the quality of environment. The valuation can be arrived from either the demand or the supply side. On the demand side, consumers insist on superior quality of environmental services like clean air, potable water, etc. Thus, valuation can be done from either of the two perspectives—how much an individual is willing to pay in order to acquire a certain quality of an environmental good, or alternatively, how much an individual is willing to pay in order to avoid a low quality of the same. Either way, a proxy for the value that a consumer places on an environmental good is obtained. Demand side valuation methods fall under two broad categories—physical linkage method and behavioural method. Physical linkage method includes measures like cost of illness (COI) approach, human capital approach and statistical value of life. The behavioural method can be classified based on two different characteristics. First, whether data are obtained from real-life observations or whether it is under a hypothetical set-up and secondly, whether the method yields monetary value directly or those have to be inferred through some indirect technique.

For the purpose of our study, we have used a modified method for capturing preferences for environmental services or goods, in this case it is arsenic safe water. We started by developing a specific application of standard microeconomic theories for measuring changes in individual's well being and used it to interpret changes in the prices and quantities of goods purchased from the market by him.

A. Das et al., *Socio-Economic Analysis of Arsenic Contamination of Groundwater in West Bengal*, India Studies in Business and Economics,
DOI 10.1007/978-981-10-0682-1_14

We assumed that individuals have well defined preferences among alternative bundles of products and services comprising both market and non-market goods. We also assumed that the agents know their preferences and there is substitutability among alternative goods in a bundle. Substitutability implies that if the individual reduces the consumption of one good, then it is possible to increase the expenditure on account of another good in the bundle, so as to leave the individual no worse off because of the change. The trade-off between different goods due to substitutability is an indicator of the values that people place on different goods. These measures of values based on substitutability can be expressed in terms of willingness to pay (WTP). We have already dealt with these in detail in the preceding chapter.

The mitigation efforts through technology deployment for arsenic contamination of groundwater were based on the assumption that the supply of safe water will automatically give rise to demand! Unfortunately, that has not been the case. That is why in rural West Bengal and Bangladesh, less than 1 % of the exposed population has access to arsenic-remediated drinking water (Gadgil et al. 2012). Field observations reveal that most of the arsenic removal systems are defunct. The performance report for domestic as well as community level plants in our survey area reflects the same sorry scenario. During our in-depth research and survey we found time and again that the actual price paid by the victims of arsenic contamination of groundwater is severe. Leaving aside only physiological, they are also economical, social, and psychological in nature almost in all of the cases. However, while the policy formulations and their implications get accounted these costs remain 'hidden'.

Take for example, Debipur village of Jalangi block. Various arsenic removal filtration systems were supplied by several organisations here. However, with eleven reported deaths in the locality and many seriously affected, it may be concluded that the safe water supply is yet to become universally available. In fact, PWSS had also reached the village in 2011. This however happened after the remarkable persistency shown by the residents. But the joy of having a safe water supply was short-lived. The water pipes have been damaged due to widening of roads in the region. The defunct connection now lies hidden somewhere under several layers of dirt and asphalt. The residents of Debipur have paid the ultimate price of development. They are unsure when the connection will be restored again.

In Debipur village, our field observations indicate, that investments made by the state as well as affected individuals have sunk. These expenditures have not yielded value for the intended beneficiaries and in fact have become burdens on the exchequer in the form of non-performing assets. In the socio economic context of Murshidabad these incremental losses have exacerbated the day-to-day struggle of the affected to stay alive. On one hand they have made unwise investments based on improper information, on the other they have been robbed of benefits of development programmes due to improper planning and execution.

They have been sucked into the vicious cycle of poverty by making continual losses on all accounts of avoidance, averting as well as travel costs. Their ownership of low resources coupled with their lessening capability to perform has ensured that the true impact of arsenic contamination of groundwater becomes intergenerational.

Death of the first affected is merely a milestone in the process of engulfing of the entire family into a trap of all-consuming poverty.

In fact, the impact of arsenic goes beyond mere resource poverty. It also has several psychosocial implications.

"Social exclusion is about more than income poverty. It is a short-hand term for what can happen when people or areas face a combination of linked problems, such as unemployment, discrimination, poor skills, low incomes, poor housing, high crime and family breakdown. These problems are linked and mutually reinforcing" (SEU 2004). In arsenic affected regions of Murshidabad these words ring absolutely true.

The hidden cost of social exclusivity has put a great stress on the society. Take for example the case of certain Mr. Jorabul Master of Khidirpara village of Domkal block. He has a large tract of land. However, he decided in favour of leaving the village rather than dying a gradual death. His son now earns his living as a daily wage labourer in the Indian state of Madhya Pradesh. Jorabul was afraid that his son may die by living in the village. He chose periodic separation with his son as a more viable option than losing him for good. He wishes his son never comes back to the village.

Mr. Lalan Mondal, a resident of the same village, belongs to another arsenic affected family. His family's financial constraints due to arsenic-induced poverty meant that he had to leave school in his teens. Subsequently, he shifted to Goa, India in search of a livelihood. Presently based out of Kerala, India he dreads the prospect of visiting his village. However, he still instils the courage and come calling for few days each year. He says, 'not staying near your loved ones is bearable. Death is not.' In several villages of Murshidabad, arsenic has forced many to leave the security of their homes. In Khidirpara more than 70 % of all households have at least one member living outside the state. The money they send home is keeping their families alive, literally. However, ironically, due to lack of earning capacity they are unable to shift their entire families out of the district. And whatever money they are sending back is exhausted on account of healthcare or in search of safe water sources and not on asset building.

From the Field 13

Any visitor to Harekrishnapur village, Khairamari GP, Jalangi block, Murshidabad is met with immense hostility by the villagers; and the source of this anger is the decade long apathy expressed by all towards their dire plight. Today, Harekrishnapur, despite being named after Krishna, the god of love and affinity, in the words of its residents is 'a laboratory dedicated to the cause of advancement of science!'

Mr. Jiauddin Saikh of this village was a trained craftsman earning respectably before the onset of arsenic-induced illness. The disease crippled his efficiency significantly. With the hope of reclaiming his life and livelihood, Jiauddin got himself admitted to SSKM hospital, Kolkata. The treatment saved his life but rendered him invalid for life. His right hand, from the elbow onwards, had to be amputated. The surgery, an expensive one by his standards, marked the beginning of his end.

Today, Mr. Saikh's wife is the mainstay of the family. Her income as a cook in a school under the midday meal scheme is helping the family scrape a living. But all is not well with her either. She had taken a substantial amount as loan, principally from neighbours, for the treatment of Jiauddin. The loan sits heavy on her mind and has resulted in bouts of depression and severe anxiety. But the most significant cause of worry for the couple is the uncertainty that looms over the future of their beloved daughter. They do not know where from the money for her education will come. They are uncertain whether they will have enough money to marry her off honourably. They are not even sure whether she will have suitable suitors now that the arsenic catastrophe of Harekrishnapur is known publicly. They do not know whether she will ever have a decent life. And as parents, they hate to have not secured her future.

But, surely he is not alone in his plight. In Harekrishnapur, stories like these are ubiquitous.

Debipur, literally the land of goddesses, ironically shares an unfortunate similarity with Harekrishnapur. We met Mr. Ashok Roy in Debipur during our visit in 2007. His plight, in more than one ways was similar to that of Mr. Saikh. Mr. Roy too was a well-to-do businessman and had the ability to invest in domestic water filters. However all popular products in the market were unable to successfully filter arsenic.

Ashok invested his faith and money on technology. However, his investment did not prevent his expenditures on treatment. He eventually succumbed to arsenic-induced illness. He is survived by his wife and two children, now living in abject poverty. His case seems like a metaphor of the government strategy to mitigate the crisis. And it reinforces our findings that investment in technology alone cannot save precious lives.

By the way, while Ashok was investing in one filter after another in search of the most efficient filtrate mechanism that may have saved his life, UNICEF promoted arsenic removal filters, available for as little as Rs. 550, lay in stacks of waste in the backyard of a certain block development officer's office!

A Catastrophe Sans Border

Abstract The crisis emanating due to the arsenic contamination of groundwater is not limited to the geographical extents of West Bengal, or for that matter India. In fact, the contamination is spread across the entire region known as the Ganga-Meghna-Brahmaputra (GMB) plains. A substantial segment of these killer plains falls within one of India's neighbouring nation: Bangladesh. The crisis at Bangladesh has been documented in-depth by international media and other stakeholders. However, there has not been any comparative study involving the worst affected Indian state of West Bengal and Bangladesh. Using secondary data for Bangladesh and primary data from our field survey, we undertake a systematic review of the crisis and its several mitigation measures. We also strive to develop a compendium of social, sociocultural, and socioeconomic effects of the same across borders.

The detection of arsenic contamination of groundwater in Bangladesh occurred in Chapai Nawabgonj, in 1993 (BGS 2001; Masrur et al. 2011). The case of Bangladesh stands out, because about 103 million people (i.e. 80 % of the rural population in Bangladesh) depend on shallow tube wells those that have been excavated at a depth of less than 150 m, thereby exposing more than half of the Bangladeshi population to high levels of arsenic and in turn made them vulnerable to arsenic ingestion-related health hazards (Nahar et al. 2008). The figures for those affected, according to a national drinking water quality survey conducted in 2009 (Flanagan et al. 2012), were found to be a staggering 20 million for those exposed to concentrations above the national standard of 50 μg/L and an even higher 45 million for those exposed to concentrations above the World Health Organisation's (WHO) guideline value of 10 μg/L! In fact, further calculations based on the findings of the same study reveal that approximately 24,000, if adhered to the WHO guideline value, and as many as 19,000, if adhered to the national standard, adult deaths occur due to chronic arsenic exposure in the country, annually. A large number of publications have studied the epidemiological issues and adverse health effects related to chronic arsenic toxicity (Datta 1976; Milton 2003; Chakraborty et al. 2004; Vahter 2009). To sum up the extent of the crisis, one may simply refer

A. Das et al., *Socio-Economic Analysis of Arsenic Contamination of Groundwater in West Bengal*, India Studies in Business and Economics,
DOI 10.1007/978-981-10-0682-1_15

to the fact that arsenic-related deaths accounted for up to 15 % of all adult deaths in affected regions of Bangladesh. These deaths, in economic terms represented an economic burden of 13 billion United States dollars (US$) in lost productivity alone over the subsequent two decades (Flanagan et al. 2012).

Apart from this tremendous loss of human resources and the resultant economic damage, the social cost of the crisis has also been unmistakeable (Hassan et al. 2005). The social implications of the crisis, unfortunately, are remarkably similar to that discovered by us in arsenic-affected regions of West Bengal. The social fallouts of the catastrophe, in Bangladesh, ranged from social ostracising, facing of rejection, being barred from social activities, at times, even by their closest of kin. Marriage was a difficult proposition for women with visible arsenicosis symptoms. Those married were struggling to survive under the constant threat of divorce. In fact, arsenic-affected women were subject to several domestic harassments and violence such as dowry, divorce, abandonment, physical distress and polygamy. Moreover, there have been studies that deduced that children who showed arsenic-induced symptoms were prevented from attending schools in an effort to keep the problem under wraps (WHO, Geneva 2000; Barkat 2004; Hadi and Parveen 2004; Hanchett 2004; Hassan et al. 2005; Ahmad et al, 2007; Nahar et al. 2008; Brinkel et al. 2009; Sarker 2010; Mahmood and Halder 2011; Sultana et al. 2012; Syed et al, 2012).

Quite evidently, the social cost of this crisis is tremendous, if not unbearable. It is equally high on both sides of the border. At both West Bengal as well as Bangladesh people have been dying due to arsenic contamination of groundwater (Table 1). They have been forced into an endless cycle of poverty. They have paid by their lives. Yet, investment to figure out the most appropriate solution to the crisis has been directed, majorly, towards technology development and deployment.

Table 1 Groundwater arsenic contamination in Bangladesh (as on survey report up to September, 2010)

Parameters	Bangladesh
Area (sq km)	147,570
Population (in million)	124.3
Total arsenic-affected districts (As > 50 μg/L)	50
Total arsenic-affected blocks/PS (As > 50 μg/L)	197
Total hand tube well water samples analysed	52,202
% of samples having arsenic > 10 μg/L	43.0
% of samples having arsenic > 50 μg/L	27.2
Total number of biological samples analysed (Hair, Nail, Urine)	11,298
Total people screened by medical group of SOES-JU	18,991
People registered with arsenical skin lesions	3762
People drinking arsenic-contaminated water > 10 μgL^{-1} (in million)	36.6
People drinking arsenic-contaminated water > 50 μgL^{-1} (in million)	22.7

Source Das et al. (2013)

Annexure 1

See Tables A.1, A.2, A.3, A.4, A.5, A.6, A.7A, A.7B, A.8, A.9, A.10, A.11 and A.13.

A. Das et al., *Socio-Economic Analysis of Arsenic Contamination of Groundwater in West Bengal*, India Studies in Business and Economics, DOI 10.1007/978-981-10-0682-1

Table A.1 Areas and population at risk (Census 2001) in the arsenic-affected areas of the State of West Bengal PHED (WB) Report

District	Total area (sq. km)	Total Population (2001)	Total No. Of Blocks	Number of affected blocks	Name of the affected blocks	Area of the affected blocks (km^2)	Population of affected blocks (in 000)	% of area affected	% of population at risk
Malda	3733	3,290,468	15	7	English Bajar	251.52	226	41.95 (4)	49.39 (3)
					Manik Chak	321.77	214		
					Kaliachal-1	105.37	311		
					Kaliachak-2	222.73	212		
					Kaliachak-3	260.12	284		
					Ratua-1	230.53	217		
					Ratua-2	173.93	161		
Sub-total						**1565.97**	**1625**		
Murshidabad	5324	5,866,569	26	19	Raninagar-1	146.93	155	64.53 (3)	68.13 (2)
					Raninagar-2	175.13	156		
					Domkal	305.19	312		
					Nawda	231.39	196		
					Jalangi	210.63	216		
					Hariharpara	253.14	222		
					Beldanga-1	168.75	259		
					Suti-1	143.68	139		
					Suti-2	111.13	213		
					Bhagwangola-1	136.1	163		
					Bhagwangola-2	175.26	130		
					Berhampur	314.19	379		
					Ragthunathganj-2	121.6	193		
					Murshidabad-Jiaganj	192.13	200		
					Farakka		220		
					Samserganj	132.74	212		
					Lalgola	84.21	268		
					Beldanga-2	184.37	210		
					Raghunathganj-1	207.93	154		
						140.91			

(continued)

Table A.1 (continued)

District	Total area (sq. km)	Total Population (2001)	Total No. Of Blocks	Number of affected blocks	Name of the affected blocks	Area of the affected blocks (km^2)	Population of affected blocks (in 000)	% of area affected	% of population at risk
Sub-total						**3435.41**	**3997**		
Nadia	3927	4,604,827	17	17	Karimpur–I	215.78	167	96.70 (1)	83.67 (1)
					Karimpur–II	224.38	192		
					Tahatta–I	249.55	218		
					Tahatta–II	172.47	134		
					Kaliganj	320.02	291		
					Nakashipara	360.94	335		
					Nabadip	97.39	122		
					Hanskhali	246.29	261		
					Krishanganj	151.6	133		
					Haringhata	170.32	208		
					Chakda	288.8	363		
					Santipur	171.41	216		
					Chapra	305.97	272		
					Ranaghat-1	145.53	207		
					Ranaghat-2	279.03	330		
					Krishnanagar-1	273.19	280		
					Krishnanagar-2	124.37	124		

(continued)

Table A.1 (continued)

District	Total area (sq. km)	Total Population (2001)	Total No. Of Blocks	Number of affected blocks	Name of the affected blocks	Area of the affected blocks (km^2)	Population of affected blocks (in 000)	% of area affected	% of population at risk
Sub-total						**3797.04**	**3853**		
North 24-Parganas	4094	8,934,286	22	19	Habra–I	117.36	188	72.06 (2)	43.12 (4)
					Habra–II	112.67	150		
					Barasat–I	104.97	238		
					Barasat–II	114.04	169		
					Deganga	202.09	276		
					Basirhat–I	111.84	147		
					Basirhat–II	127.42	194		
					Swarupnagar	215.13	226		
					Sandeshkhali–II	197.21	136		
					Baduria	179.72	248		
					Gaighata	243.3	300		
					Rajarhat	72.9	145		
					Amdanga	139.27	166		
					Bagda	233.47	220		
					Bongaon	336.7	344		
					Haroa	152.73	182		
					Hasnabad	153.07	177		
					Barrackpore–II	40.74	159		
					Barrackpore–I	95.44	189		

(continued)

Table A.1 (continued)

District	Total area (sq. km)	Total Population (2001)	Total No. Of Blocks	Number of affected blocks	Name of the affected blocks	Area of the affected blocks (km^2)	Population of affected blocks (in 000)	% of area affected	% of population at risk
Sub-total						**2950.07**	**3854**		
South 24-parganas	9960	6,906,689	29	9	Baruipur	226.16	352	12.11 (5)	28.68 (5)
					Sonarpur	120.63	167		
					Bhangar-1	153.62	204		
					Bhangar-2	162.04	207		
					Bugde Budge -2	78	173		
					Bishnupur-1	116.36	206		
					Bishnupur-2	81.71	191		
					Jainagar-1	131.01	219		
					Magrahat-2	136.93	262		
Sub-total						**1206.46**	**1981**		
Howrah	1467	4,273,099	14	2	Uluberia-2	70.47	162	11.71 (7)	7.79 (7)
					Shyampur-2	101.32	171		
Sub-total						**171.79**	**333**		
Hoogly	3147	5,041,976	18	1	Balagarh	202.15	215	6.42	4.26
Sub-total						**202.15**	**215**		
Burdwan	7024	6,895,514	31	5	Purbasthali-1	148.44	183	12.03	11.56
					Purbasthali-2	192.47	188		
					Katwa-1	168.94	152		
					Katwa-2	163.2	120		
					Kalna-2	172.17	154		
Sub-total						**845.22**	**797**		
Grand total	38676	4,581,3428	172	79		**14174.1**	**16,655**	**36.65**	**36.4**

PHED (WB) Report, 2006 (figures in the parentheses represent ranks)

Table A.2 Chronology of recommended WHO values for arsenic in drinking water

1958	First WHO International Drinking Water Standard: 0.2 mg/L (200 µg/L)
1963	WHO recommend lowering guide value to 0.05 mg/L(50 µg/L)
1974, 1984	Affirmation of 0.05 mg/L as guide value
1993	WHO provisional guideline recommends lowering guide value to 0.01 mg/L (10 µg/L)

Source WHO (2004)

Table A.3 Arsenic standards in various countries

Countries	Arsenic standard (mg/L)	Arsenic standard (µg/L)
Australia(1996)	0.007	7
Europian Union (1998)	0.01	10
Japan (1993)	0.01	10
Jordan (1991)	0.01	10
Laos (1999)	0.01	10
Mongalia (1998)	0.01	10
Namibia	0.01	10
Syria (1994)	0.01	10
USA (2001)	0.01	10
Vietnam	0.01	10
Canada	0.025	25
Baharin	0.05	50
Bangladesh (1997)	0.05	50
Bolivia (1997)	0.05	50
Chaina	0.05	50
Egypt (1995)	0.05	50
India	0.05	50
Indonesia	0.05	50
Oman	0.05	50
Mexico[a]	0.05	50
Philipines (1978)	0.05	50
Saudi Arabia	0.05	50
Srilanka (1983)	0.05	50
Vietnam (1989)	0.05	50
Zimbabwe	0.05	50
Cambodia	0.05	50
Nepal	0.05	50
LAO PDR (1999)	0.05	50
Myanmar	0.05	50
Pakistan	0.05	50

Source Ahmed (2003)
[a]Considering to lower the standard from 50 µg/L

Table A.4 Groundwater arsenic pollution around the world

Country	Source	Year of identification	Range of contamination	Population at risk	Affected areas/zones
Asia					
Bangladesh	Natural	1995	50–3200 μg/L	The population of 50 arsenic-affected districts in Bangladesh is 105 million, in comparison to the total population of the country, 120 million	50 districts (out of total 64) are now affected. The area of 50 As-affected districts in Bangladesh is 118,849 km^2 in comparison to the total area of the country, 148,393 km^2
Pakistan	Natural	2000	Arsenic concentration over the 10 ppb was detected in 32.5 % samples and more than 50 ppb was found in 10 % of the samples	In Southern part of the Punjab, an estimated population of 1.3 million is affected	Southern parts of Punjab (Multan, Bahawalpur, Rahim Yar Khan) and Central Sindh (Khairpur, Mirs and Dadu)
China	Natural and anthropogenic	First identified in Xinjiang Province in early 1980s	50–2000 μg/L	The most recent estimates show that the drinking water-type arsenicosis occurs in eight provinces/regions and the total population at risk from endemic arsenicosis from 40 counties in the provinces is 52.26 million	Xinjiang, Inner Mongolia, Shanxi, Guizhou, Ningxia, Jilin, Qinghai, Anhui Province and certain suburbs of Beijing
Thailand	Anthropogenic (mining activities)	1987	1–5000 μg/L	In Ronpibool district, most cases were of relatively mild disease, with 21.6 %, however, having very significant lesions	Ronpibool district, Nakorn Sri Thammarat Province of Thailand

(continued)

Table A.4 (continued)

Country	Source	Year of identification	Range of contamination	Population at risk	Affected areas/zones
Taiwan	Natural	1968	10–1820 μg/L	Chronic arsenicism was observed in a population of 40,421 in 37 villages, along with 7418 cases of hyper-pigmentation, 2868 of keratosis and 360 of blackfoot diseases	South-west coast of Taiwan
Nepal	Natural	Late 1990s	10–2620 μg/L	Around 0.5 million people in Terai are living at risk of arsenic poisoning. Recent studies have also reported 1.3–5.1 % of dermatosis related to arsenicosis	Terai region
Vietnam	Natural	1997	1–1050 μg/L (average concentration 430 μg/L)	High arsenic concentrations in the tube wells indicate that millions of people consuming untreated groundwater might be at considerable risk of chronic arsenic poisoning	Arsenic contamination of the Red River alluvial tract in the city of Hanoi, Mekong river delta, Ha Nam Province
Myanmar	Natural	2002	Large variations in arsenic concentrations have been found in 12 townships: Out of 793 shallow tube wells containing more than 100 μg/L arsenic, 66 % had concentration within 100–199 μg/L range; 19 % in 200–299 μg/L range; 6 % in 300–399 μg/L range; 9 % of the samples were found to contain arsenic >400 μg/L	A clinical survey of arsenicosis patients was undertaken among 6976 persons from 79 worst-affected villages in three townships of Ayeyarwady division. There have been 21 cases of dermal changes, 8 suspected cases of arsenicosis based on pigmentary changes	Bago, Ayeyarwady, Rakhine divisions, Mandalay divisions

(continued)

Table A.4 (continued)

Country	Source	Year of identification	Range of contamination	Population at risk	Affected areas/zones
Japan	Natural and anthropogenic	1994	1–293 μg/L	As of 1995, there were 153 patients in Toroku and 64 in Matsuo who were recognised by the government as suffering from chronic As poisoning	Southern region of Fukuoka Prefecture, Sendai,Tkasuki, Kuamoto,Nakajo, Toroku & Matsuo
Cambodia	Natural	1999/2000	Concentration in soil 8–16 ppm (dry weight)	First clinical survey was carried out in Kandal Province in October–November 2003. 7817 people from 12 villages have been examined for arsenical skin lesions. No case of arsenicosis was identified during the survey	Kandal Province
Afghanistan	Natural	Not available	10–500 μg/L	The estimated population that is potentially at risk is around 500,000	Logar Province, Ghazni metropolitan area
Iran	Natural	1981	Arsenic in water supplies exceeding 1000 μg/L with an average concentration of 290 μg/L. Measurements have shown levels in some water resources as high as 1480 μg/L	Chronic poisoning manifestations such as skin lesions (keratosis, pigmentation) and even amputation due to gangrene were reported in these villages	Kurdistan, a western province of Iran

(continued)

Table A.4 (continued)

Country	Source	Year of identification	Range of contamination	Population at risk	Affected areas/zones
Europe					
Greece	Anthropogenic Source (lead mining and smelting)	Not available	Arsenic concentrations in garden soil and house dusts around the ancient lead mining and smelting site at Lavrion ranged up to 14,800 and 3800 mg/kg, respectively	Local community of lead mining and smelting site of Lavrion	Lavrion
Bulgaria	Anthropogenic Source (copper smelting)	Not available	The water samples from Topolnita River had As ranging from 750 to 1500 μg/L, as revealed in a survey from 1987 to 1990	Exposed population 32,000	Srednogorie, the largest metallurgical centre in Bulgaria
Hungary	Natural	1941–1983	60–4000 μg/L	A few thousand people showed symptoms of arsenic poisoning	The plain where arsenic was found about 1,10,000 sq. km.
Romania	Natural	Not available	0–176 μg/L	The estimates have shown that about 36,000 people were exposed to arsenic concentrations in drinking water ranging from 11 to 48 μg/L and about 14,000 inhabitants were exposed to arsenic concentrations above 50 μg/L	Northwestern parts of Transylvania

(continued)

Table A.4 (continued)

Country	Source	Year of identification	Range of contamination	Population at risk	Affected areas/zones
Switzerland	Natural	Not available	10–170 μg/L	Not available	Jura Mountains and in the Alps
United Kingdom	Anthropogenic source (mining activities, a British plant manufactured sodium arsenite from 1910 to 1943, As from a steel foundry)	Not available	From limited available data, three private supplies of those tested in Cornwall had arsenic levels above the 5 μg/L, detection limit and contained 11, 60, 80 μg/L. The soil As concentrations in Armadale were generally higher (52–64 μg/g) than those in the control uncontaminated samples	75 deaths were reported among factory workers, of which 22 (29 %) were due to cancer. In Armadale, a town in central Scotland with a population of 7000, the standardsed mortality ratio (SMR) for respiratory cancer was high from 1969 to 1973 due to exposure to arsenic from a steel foundry located in that area	Cornwall, Armadale
North America					
USA	Natural and anthropogenic sources (copper smelting and gold smelting)	Not available	Arsenic concentration range varies from one county to another county. In milliard country (Utah), arsenic concentration ranges from 180 to 210 μg/L in Western Oregon, arsenic concentration varies from 50 to 1700 μg/L in Lessen country (California) and arsenic in the range 50–1400 μg/L was reported	Mortality rates from ischemic heart disease significantly increased among arsenic-exposed workers of copper smelter. The initial 1938–1963 mortality analysis of workers at the copper smelter at Anaconda demonstrated a more than threefold excess respiratory cancer ratio, with excess risk as high as eightfold among heavily exposed men who had worked there eight years or more	Millard County, Utah; Western Oregon; Lessen County, California; Fallon, Nevada; New Hampshire; Fairbanks, Alaska

(continued)

Table A.4 (continued)

Country	Source	Year of identification	Range of contamination	Population at risk	Affected areas/zones
Mexico	Natural and anthropogenic (mining activities)	1963–1983	21–1070 μg/L	Exposure to arsenic associated with mining activities arises from drinking water, soil, dust and the estimated exposed 600,000	Lagunera Region, in the central part of North Mexico, city of Torreon, Silesia, San Luis de Potosi, in central Mexico, State of Hidago in the Zimapan valley. The area affected is 32,000 sq. km
South America					
Argentina	Natural	1917	100–2000 μg/L in some countries of Corboda, arsenic concentration ranges from 100 to 2000 μg/L in a study of the Carcarana river basin (Corboda and anta Fe Provinces), observed arsenic concentrations of less than 10–720 μg/L (average 201 μg/L), while less than 4–5280 μg/L arsenic concentration in groundwater in La Pampa Province	2,750,000	Cordoba, Salta, La Pampa, Santa Fe, Tucuman, Santiago del Estero, san Luis and parts of Buenos Aires (an estimated area of 165000 sq. km.)

(continued)

Table A.4 (continued)

Country	Source	Year of identification	Range of contamination	Population at risk	Affected areas/zones
Chile	Natural and anthropogenic	1962	40–860 μg/L	Population exposed in 1996 to specific levels of arsenic in the drinking water in Chile:7 million inhabitants (53.3 %) were exposed to less than 10 μg/L; 5.5 million (41.9 %) were exposed to 10–30 μg/L; 450,000 inhabitants(3.4 %) were exposed to 50–60 μg/L; and 1500 (0.01 %) were exposed to 600–800 μg/L	Antofagasta, Calama and Tocopilla
Other countries					
Africa (Ghana)	Natural	Not available	1–141 μg/L	100,000	Obuasi area in the Ashanti region and the Bolgatanga area of the upper East region
Australia	Anthropogenic source (mining activities and pesticide uses)	Not available	Varied from <1 to 300,000 μg/L in groundwater	Arsenic poisoning among the workers	Victoria, mostly around the gold mining regions

Source Ahmed (2003), Rahaman et al. (2006), Mukherjee et. al. (2006)

Table A.5 Water quality problems across states in India

Parameter	Maximum permissible limit (mg/L)	Affected areas
Fluoride	1.5	Andhra Pradesh, Assam, Bihar, Chattisgarh, Gujarat, Jharkhand, Karnataka, Orissa, Madhya Pradesh, Maharashtra, Punjab, Rajashthan, Uttar Pradesh, Tamil Nadu, Tripura, West Bengal
Arsenic	0.05	Assam, Bihar, Chattisgarh, Jharkhand, Tripura, Uttar Pradesh, West Bengal
Iron	1	Arunachal Pradesh,Assam, Bihar, Kerala, Jharkhand, Karnataka, Jammu and Kashmir, Manipur, Meghalaya, Mizoram, Maharashtra, Madhya Pradesh, Nagaland, Orissa, Punjab, Rajasthan, Sikkim, Tripura, Tamil Nadu, Uttra Pradesh, West Bengal, Andaman & Nicobar (A&N) Islands, Pondicherry
Salinity	2000	Andhra Pradesh, Chattisgarh, Gujarat, Haryana, Kerala, Madhya Pradesh, Maharashtra, Orissa, Punjab, Rajasthan, Tamil Nadu, Pondicherry, Uttar Pradesh, West Bengal
Nitrate	100	Bihar, Gujarat, Karnataka, Kerala, Madhya Pradesh, Maharashtra, Punjab, Rajasthan, Tamil Nadu, Uttar Pradesh
Heavy metals	Cadmium—0.01 Zinc—15 Mercury—0.001	Gujarat, Delhi, Andhra Pradesh, Haryana, Kerala

Source Water Aid (2008) www.ddws.nic.in. Accessed on Jan 2009)

Table A.6 Water quality in India—extent of the problem

Types of water quality problem	Number of affected habitations
Excess iron	118,088
Excess fluoride	31,306
Other problems	25,092
Excess salinity	23,495
Excess nitrate	13,958
Excess arsenic	5029
Total	216,968

Source www.ddws.nic.in. Accessed on 21st June 2009

Table A.7A Progressive spread of arsenic contamination of groundwater in West Bengal (study conducted by PHED, WB)

Year	No. of affected district	No. of affected blocks	No. of affected village	No. of affected habitations	No. of affected municipalities
May 1993	7	34	78	149	3
September 1995	7	56	388	757	9
December 1997	8	61	1302	3049	9
December 1998	8	65	1312	3067	9
December 1999	8	67	1550	3365	11
December 2001	8	75	2065	4244	11
December 2002	8	75	2579	4973	11
March 2006	8	79	3235	6623	11

Source PHED, WB (2006)

Table A.7B Progressive spread of arsenic contamination of groundwater in West Bengal (study conducted by SOES, JU)

Year	Number of affected districts	Number of affected blocks	Number of affected villages
1995	6	37	405
1998	8	61	863
June 2002	9	81	3200
December 2005	12	111	3417

Source SOES (JU) Report (various years)

Table A.8 Long period averages of net area irrigated by different sources, 1950–2007

Years	Canals	Tanks water	Total surface	Tube wells	Other wells	Total groundwater	Others (incl. both sw/gw)	NIA
1950–51 to 1964–65	42	18	60	3	29	32	8	100
1965–66 to 1979–80	40	12	52	16	24	40	8	100
1980–81 to 1994–95	37	7	44	29	21	50	6	100
1995–96 to 2006–07	28	4	32	39	21	60	8	100

Source Indian Agricultural Statistics (various issues)

Table A.9 Number of government-owned tube wells to supply drinking water (2007)

Districts	Distribution of Installed government-owned tube wells	Rank of the districts with respect to number of installed
Murshidabad	35,995	1
Burdwan	32,641	2
Paschim Madinipur	31,641	3
South 24-Parganas	29,849	4
Nadia	25,701	5
Hoogly	24,464	6
Koochbehar	23,035	7
Utta Dinajpur	21,762	8
North 24-Parganas	21,451	9
Purba Medinipur	20,491	10
Bankura	19,832	11
Birbhum	18,436	12
Malda	15,330	13
Jalpaiguri	15,084	14
Howrah	14,013	15
Dakshin Dinajpur	13,139	16
Purulia	11,915	17
Darjeeling	5626	18
Total	380,405	

Source Various reports (published /unpublished from arsenic task force and PHED)

Table A.10 Different concentration levels of arsenic in the groundwater samples, tested by PHED (JPOA) and SOES (JU)

West Bengal	Study conducted by			
	PHED (JPOA)		SOES(JU)	
	Number	Percentage	Number	Percentage
No. of samples	1,32,262		1,40,150	
<10 μg/L	55,568	42.03	72,681	51.86
10–50 μg/L	42,877	32.42	33,993	24.25
>50 μg/L	33,755	25.52	33,476	23.89
>WHO level	76,632	57.94	67,469	48.14

Source SOES (JU) and PHED (various years)

Table A.11 Social manifestation of arsenicosis

Social Manifestation	Percentages
(1) Arsenic patients are now burden of the family	16
(2) Facing serious financial constraint	20
(3) Arsenic patients are treated as untouchable	8
(4) Men and Women are finding it difficult to get married	12
(5) Rate of dessertation of partner having such ailments is also rising among married couples	4
(6) Social Instability	8
(7) Suffering from depression	12
(8) Parents are suffering from depression thinking of social exclusion due to arsenic threat	4
(9) Shrinking of job opportunities as few employers are keen to hire patients of arsenic poisoning	4
(10) Self-employed are finding it difficult to keep their business running due to physical disabilities	4
(11) There is migration due to arsenic threat	8
(12) Arsenic patients often try committing suicide	4
(13) High indebtedness to meet treatment cost for arsenic ailments	20
(14) Assets sold off to meet health expenditures caused by arsenicosis	4
(15) Problem of selling cultivated products	4
(16) Physical weakness and loss of earning capacity	20
(17) School of drop-out and child labour	12
(18) Deaths occurred due to arsenicosis	12
(19) Patients prefer to remain unidentified	4
(20) Loss of faith	4
(21) Role of the panchayat is not satisfactory	8
(22) No alternative source of drinking water	4
(23) We are poor, it is difficult to get nutritious food	4
(24) Financial constraint	4

(continued)

Table A.11 (continued)

Social Manifestation	Percentages
(25) Spending money on treatment is a waste	4
(26) Inadequate stand posts in the village due to which long queue used to form	4
(27) No government and non-governmental initiatives have came up as a permanent solution	4
(28) Nothing has been done	4
(29) Lack of proper maintenance has rendered plants defunct	4
(30) Lost faith and confused	4
(31) Government officials promised to provide arsenic safe water to the villagers, but were not realised	4
(32) No youths in the village	4
(33) Forced to leave the village	4
(34) Not able to educate sons	8
(35) Feeling hidden hatred brewed	4
(36) Relatives discontinued visiting arsenic-affected villages	4
(37) Occupational disadvantages	4
(38) Use razor blade to scrape off rash caused by arsenic	4
(39) High amount of dowry	4
(40) Not taking any medicine for financial constraint	4
(41) Medical delivery system is not good	8
(42) Body itches burn all the time. Cannot go outside during the day time	16
(43) Physical weakness and cannot do hard work	12

Source Field survey

Table A.12 District-wise arsenic mitigation schemes and population coverage (Upto March 2007)

	Districts																	
	Mushidabad		Nadia		Howrah		Hoogly		Burdwan		South 24-Pgs		North 24-Pgs		Malda		Total	
	No.	Pop. cov.	No.	Pop. cov.	No.	Pop. cov.	No.	Pop. cov.	No.	Pop. cov.	No.	Pop. cov.	No	Pop. cov.	No	Pop. cov.	No	Pop. cov.
Short-term measures																		
(1) Ring well	114	0.285	6	0.015	0	0	0	0	0	0	0	0	0	0	46	0.115	166	0.415
(2) New hand fitted tube wells at deeper aquifer.	2432	6.08	2113	5.28	20	0.05	33	0.083	52	0.13	694	1.74	2450	6.13	243	0.608	8037	20.09
Medium-term measures																		
(1) Arsenic treatment unit (ATU) with existing hand pump fitted tube wells	1131	2.83	272	0.68	0	0	0	0	0	0	100	0.25	702	1.76	191	0.48	2396	5.99
(2) Arsenic removal plants for existing groundwater-based pipe water supply schemes(PWSS)	0	0	0	0	0	0	0	0	0	0	0	0	10	1.5	2	0.4	12	1.9
(3)New big diameter deeper aquifer tube wells for existing PWSS	2	0.3	0	0	0	0	0	0	0	0	0	0	6	0.9	0	0	8	1.2
(4) New groundwater-based PWSS	46	9.84	59 +1[a]	11.51	6	1.27	6	1.20	19	1.82	3	0.60	89	11.36	10	4.59	239	42.19
Long-term measures																		
Surface water-based PWSS	0	0	0	0	0	0	0	0	0	0	1	16.28	1	2.59	1	7.71	3	26.58
Total	3725	19.34	2451	17.49	26	1.32	39	1.28	71	1.95	798	18.87	3258	24.24	493	13.90	10861	98.36

[a]Nadia-Murshidabad groundwater-based PWSS is considered in Nadia district. Since out of total 28 covered villages, 24 is under Nadia district. But to obtain the coverage population we consider the respective district coverage (Murshidabad = 31,169 in 4 villages, Nadia = 80,994 in 24 villages)

Source PHED Report (various years)

Technical Appendix: Threats to Sustainability

Mitigation Efforts

Demand for government action to provide arsenic safe water to attain better health standards for people has been expressed through various forms of public grievances, coverage in mass media and social conflicts between local inhabitants and governments. The locals have staged mass deputation to the government officials, adhered to other means of protesting such as hunger strike, rally, polio boycott, etc. (Das 2006, 2009). Media has occasionally reflected on the morbidity and mortality incidents due to the arsenic contamination of groundwater. It has also resorted to editorials, expert interviews and other means of reportage to draw Government's attention and raise public awareness about the health impacts of consumption of arsenic contaminated water. These emphasised the need for special action towards public provision of safe water. Researchers started informing the government and raising public awareness several years ago about the possible health impacts of arsenic problem in different parts of West Bengal. However, economists have started conducting major research on the economic valuation of arsenic-related health hazards for West Bengal and Bangladesh only recently (Roy 2008; Roy and Das 2009; Khan 2007; Das 2012).

The following schematic representation narrates the desirable properties of a sustainable mitigation effort. The scheme as shown in Fig. A.1 needs to be implemented holistically. For assessment of mitigation efforts to provide 'access to arsenic safe water' for households in arsenic-affected areas, one needs to examine the following:

1. The mitigation effort matches with the nature and scale of the problem.
2. Drinking water service system could be created through appropriate asset building with self-sustaining appropriate long-term financing mechanism.

Mitigation efforts are not costless so it is important to strategize the efforts so that the twin objectives of least-cost service delivery and longer term sustainable access to arsenic safe water can be provided and health impact is alleviated in the short run and prevented in the longer run. While analysing the problem, it has been shown by experts that ideally long-term solution to the problem lies in prevention rather than in cure. Prevention can happen only through making provision of access to arsenic safe water to rural households. Given the fact that most commonly used groundwater aquifer is contaminated with arsenic, the mitigation solutions can be the following:

1. To provide drinking water from any surface water source, river, ponds, etc., after proper treatment;
2. Regulated withdrawal of arsenic contaminated groundwater and application of appropriate technology for arsenic removal through public investment as the right to aquifer is hard to define and cannot be defined for private ownership;
3. Regulated withdrawal of groundwater from deeper aquifer which has been proven to be a source of safe drinking water;

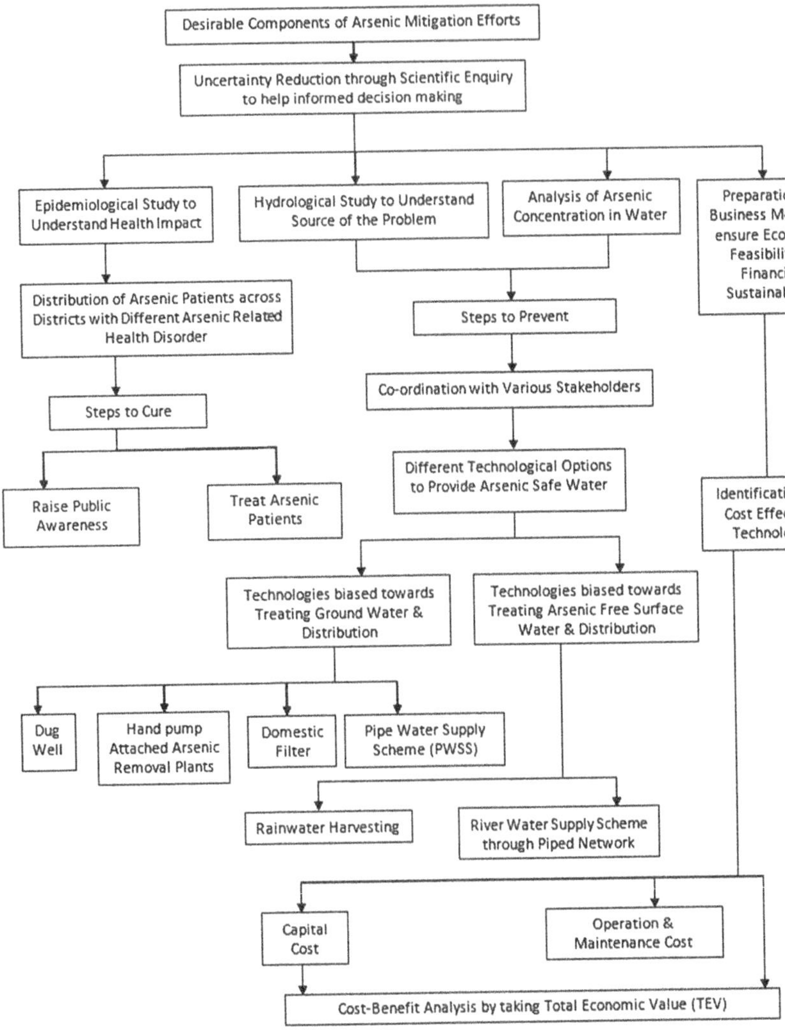

Fig. A.1 Schematic diagram

4. Either of the solutions in 1–3 needs to be provided with longer term objective through public policy/investment/regulation and end-user participation in cost sharing to ensure financial sustainability.

So the goal of any mitigation effort in West Bengal needs to be 'sustained access to arsenic safe drinking water' through public policy or intervention. Sustained access means inter-temporal access security. This can happen if any mitigation effort is undertaken:

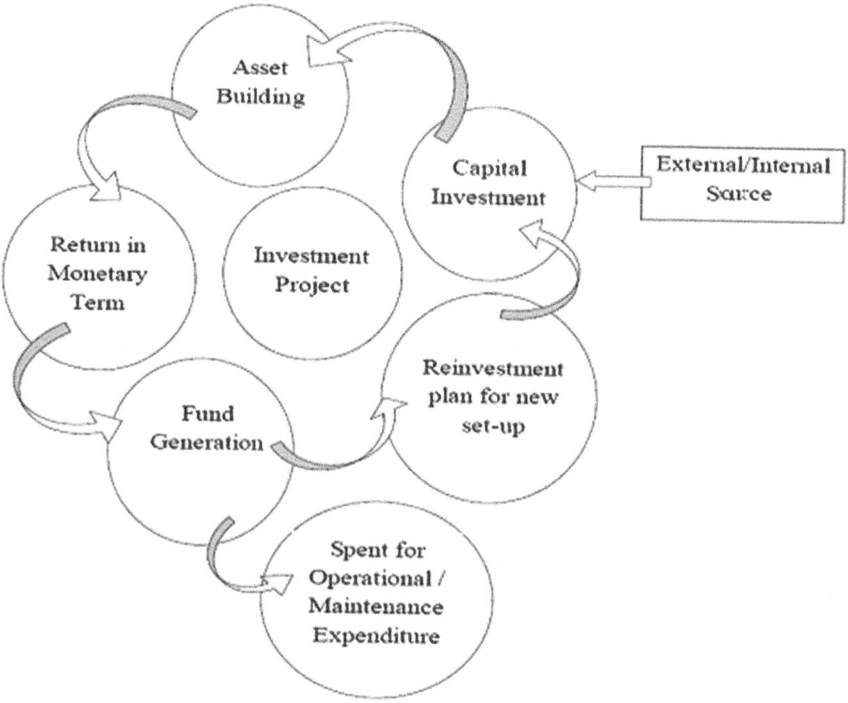

Fig. A.2 Investment model

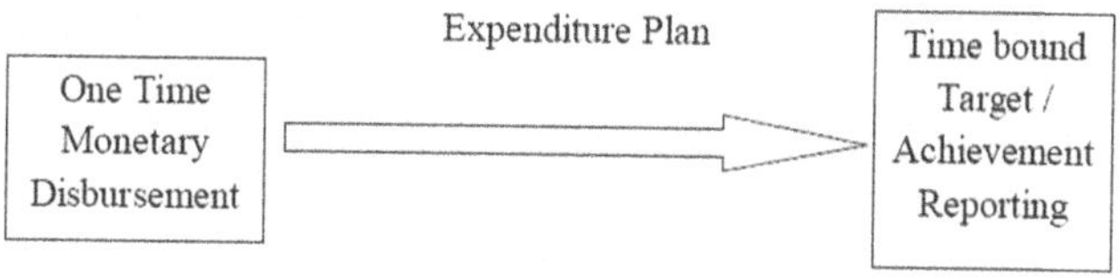

Fig. A.3 Expenditure model

1. As an investment project which can generate its own return for reinvestment in future for water supply infrastructure building and maintenance;
2. With the goal of achieving inter-temporal source security, i.e. non-declining water quality and quality management;
3. Arsenic safe water would mean additional investment in arsenic removal if source water is contaminated and it also needs to be on a return guarantee investment project mode. We assume any mitigation effort to secure for rural households 'sustainable access to arsenic safe drinking water' must be designed from the beginning as an 'investment model' (Fig. A.2) and not an 'expenditure model' (Fig. A.3).

A prerequisite for effective preventive action design through 'sustainable access to arsenic safe water' model is the availability of scientific information on the extent of the risk/exposure. Integrated water resource management (IWRM) can provide the necessary preventive mechanism and solution to the problem through these following steps:

(a) Compilation and dissemination of information on arsenic safe water source;
(b) Creation of appropriate infrastructure to deliver arsenic safe water supply service system to ensure access to end-user point;
(c) Responsible formal participation of end users in arsenic safe water supply scheme to enhance access and prevent arsenicosis.

Data

To make assessment of mitigation efforts using Fig. A.1, we need information on components of all the boxes. The quantitative and qualitative data have been collected from both primary and secondary sources. Government reports, media coverage on arsenic victims, research reports and articles have been utilised as a secondary source. To identify arsenic-affected areas as well as arsenic concentration level in the tube-well water samples, we rely completely on the data source published by PHED (WB) and SOES (JU). Since secondary information available is inadequate on household's perception on value of arsenic-free water, we have conducted a survey of households to collect the primary data for the necessary information. This data has been collected through a questionnaire-based field survey through a direct interview method. A number of focus group discussions and case studies were also conducted in each village in order to get more qualitative data. Group discussions were conducted to evaluate the socio-economic impact of arsenic contamination on the rural society. Table A.13 shows the various data requirements and sources of information used.

Arsenic Mitigation Efforts in Rural West Bengal: Performance Evaluation

Actions undertaken in the last two decades towards technology development to provide arsenic safe water in rural West Bengal have produced mixed outcomes. Claims by funding agencies, implementing agencies, show that problem has been adequately addressed and rightly approached, while local people, civil societies, experts, mass media and NGOs argue that given the enormity of scale of the problem nothing has been done on a sustained manner and the ill effects of the menace are on the rise. It is important to understand the logic behind these two claims objectively. It is essential to come up with a criterion for performance

Table A.13 Data source

Sl. No.	Components/ variables	Source of information[a]
(A)	**Uncertainty Reduction through Scientific Enquiry to enable better informed decision making**	
1.	Epidemiology study to understand health impact	(1) Research publication in various peer-reviewed journals (2) Field visit with specialised doctors and School of Environmental Studies (SOES) team
2.	Hydrological study to understand source of the problem	(1) Research publication in various peer-reviewed journals. (2) Personal communication with central groundwater water development (CGWB) at Kolkata
3.	Testing of arsenic concentration in water	(1) School of Environmental Studies (SOES), Jadavpur University, (2) Public Health Engineering Department (PHED), Government of West Bengal (3) Bengal Engineering College, Shibpur (4) Arsenic Task Force (ATF), Government of West Bengal (5) State Water Investigation Department (SWID), Government of West Bengal (6) UNICEF (7) local NGOs at Murshidabad district
4.	Preparation of business model/investment model to ensure economic feasibility and financial sustainability	Official documents on past financial allocation
(B)	**Steps to Cure**	
1.	Raise public awareness	(1) Director of Community and capacity development unit (CCDU), (2) local NGOs at Murshidabad district, (3) field visits at Murshidabad district
2.	Treat arsenic patients	(1) Doctors of STM, SSKM hospital at Kolkata, (2) local doctors at Murshidabad district, (3) visiting SasthyaBhawan, Government of West Bengal and (4) research publication by various peered reviewed journals
(C)	**Steps to Prevent**	
1.	Coordination with various stakeholders	(1) Public Health Engineering Department (PHED, Government of West Bengal, (2) Arsenic Task Force (ATF), Government of West Bengal, (3) All India Institute of Hygiene and Public Health (AIIH & PH), (4) UNICEF officials and (5) local NGOs (Sanitary Marts[b]) at Murshidabad district
2.	Different technological options to provide arsenic safe water	
(D)	**Identification of cost-effective technology**	
1.	Cost-benefit analysis by taking total economic value	Secondary sources and primary survey

[a]We also considered reports on arsenic issue published in some reputed newspapers as our data source

[b]Sanitary marts (SM) are rural production centres and retail outlets that manufacture and market low-cost hardware and provide services. This is initiated under the Government of India's Total Sanitation Campaign (TSC). The concept of production centres and SMs was first tried out in an Intensive Sanitation Project in West Bengal's Midnapur district in the 1990s. The state government implemented the project with the assistance of a local NGO, Ram Krishna Mission Lok Shiksha Parishad (RKMLSP), and UNICEF. The rural sanitary mart typically sells hardware for constructing rural toilets and bathing/washing areas. Other items relating to home and personal hygiene like long-handled ladles for drinking water pots, brooms, brushes, water filters, soaps and disinfectants are also included in its inventory. Under the TSC programme, NGOs with suitable experience and Panchayats are eligible for funding up to Rs. 3.5 lakhs for setting up and running of SMs that, in time, should become self-sustaining

evaluation of arsenic mitigation projects. This is needed to strengthen the implementation and monitoring and evaluation of the development programmes and also to reduce social conflicts over differences in claims.

Development project implementation and project outcome assessments are generic problems. It may be useful to start from the premise that project implementation and monitoring are both a science and an art in itself. This emphasises the importance of the software part of any hardware project be it a centralised water supply system or decentralised standalone systems in arsenic prone or safe areas. It is an 'art of operationalization' of a project conceived and designed at the top for local, community-level implementation in variety of socio-economic realities. Project development and implementation are two different sets of skills but need to be addressed holistically and in an integrated manner. The first encompasses technology and investment need assessment, and the latter addresses local demand and local social circumstances. A project where technology in the engineering sense or hardware is used without being appropriately packaged with software for coordination is bound to fail due to coordination failure. So how a project is conceived through its lifetime and monitored is the art of operationalization and needs equal weight as any hardware component. Most of the development projects are successful up to the stage of technology purchase and sometimes till deployment stage but failure is almost evident. Such failures and examples of unsustainable interventions in turn reduce confidence in the hardware thus jeopardising the self-sustenance of development projects. If engineering and technology is hardware of a development project, socio-economic welfare assessment, operationalization and sustainability strategy are its software. As mentioned previously the need is an investment project and not an expenditure one.

Mitigation efforts by the government started officially from 1983 through setting up of first technical committee. We make an assessment of technological options adopted. Such economic assessments are absent in the literature. Whatever literature exists focuses on severity of health impact (Saha and Chakraborti 2001; Guha Mazumder et. al. 1998, 1999, 2000), increasing spread of the problem (Chakraborti 2009), technologies available and technology installation projects and head count of beneficiaries (PHED 2006, 2007). None of the studies provide a comprehensive evaluation of what has been achieved in terms of sustenance of economic gain and welfare of the beneficiaries. This can provide guidelines on how to improve upon the performance after projects. It is important to understand how a problem is conceived by implementing agency as in the very conception lies the directions of final outcome path. Our goal is to evaluate implementation strategy of government action.

Objective is to understand what can be suggested to strengthen the implementation strategy. As framework of evaluation our primary indicator is sustainability of provisioning access to arsenic safe water and secondary indicators are nature of decision: systems or partial, investment orientation or expenditure orientation.

Performance Evaluation: BAU Criteria Versus Alternative Criteria

Conceived as expenditure projects with supply-side orientation, projects are evaluated by funding agencies, implementing agencies, government reports by monetary resource mobilisation, i.e. amount of expenditure, number of installations and head count of beneficiaries. So the current business as usual (BAU) evaluation criteria are biased and ignore the outcome evaluation. It is like accounting for inputs used in production process without accounting for the total output. In the next subsection, we provide the BAU evaluation criteria-based claims followed by a section providing alternative criteria that can answer the question that we started with: why there is gap in claims by various stakeholders?

BAU Criteria

Head count and installation counts of a standard expenditure projects are reported to explain the success story of implementation programmes. Up to April 2007, total 2396 ARPs have been set up under arsenic treatment unit-hand pump (ATU-HP) scheme and fitted to the tube wells affected by arsenic at an average price of US $1500/unit, in mainly five out of nine severely arsenic-affected districts of West Bengal (PHED 2007). Up to April 2007, total 166 dugwells, 8037 hand fitted tube wells at deeper aquifer, 259 groundwater-based PWSS and 3 surface water (river water)-based PWSS have been set up in West Bengal for supplying arsenic safe water. Through different short-term, medium-term and long-term measures, West Bengal government is now able to supply 9.84 million people (59 % of the total affected population) up to April, 2007.

A second criterion for expenditure project evaluation is goal of fund raising and documents do show positive growth. Considering the gravity of the situation, the state government prepared a project estimated at Rs. 0.1082 crore which was sanctioned by Government of India (GOI) under the aegis of Technology Mission in May, 1988 to study the nature, extent and cause of arsenic pollution huge amount of investments have already been made to supply arsenic safe water in the affected rural areas of West Bengal. Total investment for implementation of water supply schemes in arsenic-affected blocks in West Bengal up to 31 March, 2007 is Rs 876.75 crores.

Alternative Criteria: System Approach and Investment Project Evaluation Criteria

Art of implementation lies in formulation of a development project with both supply- and demand-side parameters incorporated. While supply side is reflected in costs incurred and inputs engaged in production process, demand side reflects the return that can be generated from the expenditure made. So BAU criteria of cost assessment only cannot be used to justify an investment unless return is reported as well.

We propose efficiency criteria as evaluation criterion where efficiency is defined as

$$e = y^*/y \tag{A.1}$$

y^* is the optimum benefit–cost ratio. This can be dynamic depending on the technology basket. So theoretically it will need to be revised like any benchmark value. But we can avoid that for all practical purposes and if we get value of 'e' as lesser than one then we can revise the value with the new value. Now if $e = 1$ then the project can be identified as most efficient, and if it is less than one then it is efficient and magnitude can reflect the level of relative efficiency.

For arsenic mitigation projects how do we arrive at y^*? To obtain y^* we need output/return of the project per rupee value of cost, i.e. benefit–cost ratio. The return can be judged by human welfare implication of health benefit generated per investment cost incurred. Unless monetised they cannot be compared. Epidemiological studies which are large in number cannot provide any monetary value of benefit generated. Health production function used in environmental economics literature (just to name a few relevant ones in this context (Freeman 1993; Roy 2008)) can generate monetary value of welfare gain. Such rigorous health production function study (Roy 2008) in the context of West Bengal shows reduction of arsenic concentration by 1 μg per litre, which can generate benefit of Rs. 49 ($1) for 100 households per month. This parameter of health benefit can be used to estimate return from any arsenic removal project using local concentration level, population in command area, etc. Using this methodology we estimate the total welfare gain from the arsenic mitigation projects for affected districts also (see Table A.14).

It is important to note that for arsenic mitigation project y^* can be taken as 1.6 based on our estimates[1]. Any project that can generate Rs. 1.6 worth of human welfare gain from one rupee worth of mitigation investment can be taken as bench mark project or most efficient project.

This objective efficiency-based criterion gives a handle on the achievement level and real progress. BAU evaluation criteria if replaced by the alternative efficiency criterion the project monitoring will be easy and performance will improve as there will be a benchmark target for each to achieve. This will justify the claims of all various stakeholders.

[1]It needs further refinement based on more accurate information across all districts. But can be taken as a conservative number for now.

Table A.14 District-wise estimated maximum welfare gain

Districts	Total Population (2001)	Population at Risk (As per PHED Report)	No. of Household at Risk (As per PHED Report)	Average Arsenic Concentration*	Estimated Welfare Gain for the District per Month* (in million)	Estimated Welfare Gain for the District per Year* (in million)
Murshidabad	5866569	1882000	376400	240	35	420
Nadia	4604827	1719000	343800	184	23	276
Howrah	4273099	133000	26600	147	1	12
Hoogly	5041976	129000	25800	124	0.9	10.8
Burdwan	6895514	192000	38400	181	2.5	30
South 24-Parganas	6906689	1981000	396200	413	70.50	846
North 24-Parganas	8934286	2413000	482600	244	45.88	551
Malda	3290468	1387000	277400	260	28.54	343
West Bengal		9836000	1967200	*259*	201.5	2418

Source PHED Report (2008)
*Authors' estimates

Action Programmes: Need for System Approach

An implementation strategy can be successful only if it is holistic. An approach is holistic if it adopts a 'Systems Approach'. System approach here implies both supply-side management of arsenic-free water supply service and demand-side management to ensure sustainability of the service. The whole approach of past implementation strategy has been partial in nature. It has been governed by the philosophy that '*supply creates its own demand*". So demand constraint was not in the sight of the programmes and participation was taken as automatic from end users/beneficiaries. Demand from end user has been looked upon as a passive and autonomous factor. So incentive design that triggers demand has never been in the agenda of technology deployment programme. Administratively, Public Health Engineering Department (PHED), Government of West Bengal (GoWB), has the sole responsibility of supplying drinking water in rural areas. Engineering approach focused on monitoring of capital investment and technology deployment. Top-down approach is biased by supply-side management. Technology deployment strategy has recommended an in-depth scientific investigation and information building as the first step.

Sustainability of any programme depends on how the programme is conceived for implementation. If it is conceived as investment project with long-term/perpetuity goal, then the aim will be to generate return out of the investment or plan the initial grant endowment as installation funds as well as maintenance fund to self-sustain the project. If it is an expenditure project, then it is designed as one-time investment plan. For expenditure projects, the goal is to prepare plan for spending the initial fund generated or received to achieve target installations, reach out to target beneficiaries. So head count or installation counts are the monitorable indicators. Lifetime of the projects beyond project grant period is not the goal or evaluation criterion for expenditure projects. However, problem with expenditure projects is that they do not ensure sustainability, whereas investment projects have built-in sustainability if planned correctly.

The operationalization of programmes adopted so far can be categorised as one of expenditure projects and not investment projects within the defined framework. On the basis of the recommendations of different committees, investigations and preparation of detailed work plan to acquire funds and spend them to achieve target number of installations to reach out a target number of beneficiaries have been managed the Government of West Bengal.

There has been quite systematic scientific approach to provide arsenic safe water in West Bengal. There have been water testing laboratories, arsenic removal technologies installed, involvement of government and non-government actors working towards mitigation of the problem. Investment has also been put in place. But evaluation of the attempts shows that approach to the problem has been quite scientific but implementation strategy has so far been very weak to yield any sustainable benefit out of the strategy. After a decade also, it seems that no self-sustaining financing process has developed nor the demand for arsenic safe water supply technology has made any permanent position. The reasons need to be looked into the weak institutional and economic incentive design which makes the software that drives the hardware in reality. In the absence of economic incentive-based software to drive the hardware of technologies, testing laboratories could not revolutionarise the water and health sector. So any future action needs strong economic incentive design as supplementary to technical intervention.

The approach so far taken to tackle the situation in West Bengal is dominated by technological solution. Easy money flow and easy technology flow were there but the management and economic aspects were silent and hence can be considered as an indicator of failure. Cost-effectiveness, economic valuation never figured in the agenda. As a result, financial sustainability, technical sustainability and participatory approaches never come in forward. Demand management had never been the goal. Supply management which dominated was also not planned with long-term vision or within cost-effectiveness.

Critique

Actions in the past decade have systematically approached the problem of arsenic contamination of ground water. The focus has been to address the problem of water supply sector. Search for arsenic safe drinking water supply has dominated the actions. Scientific approach to the problem has been successful in setting up of arsenic testing laboratories accessible to end users. Action plans have been implemented through public–private partnership as top-down prescription without any local innovation. Knowledge generation followed by infrastructure building goes beyond testing facility creation to real technology deployment to provide arsenic safe water. Investment flow towards hardware, i.e. deployment of arsenic removal technology and supply of arsenic safe water to rural West Bengal, has led to expenditure of financial resources into the technology purchase. Going by the reports and statistics, it may appear that arsenic menace is going to be over through technology deployment. The current research has shown that past efforts to solve the problem lived the project lifetime and service have not been always sustained at end-user point through maintenance of the technologies. Governance has been overly biased towards technology deployment and not technology sustenance through incentive design and creation of demand. Demand-side management has been completely ignored so despite investment in technology procurement there is no market or non-market-based institutional arrangement to generate consumer's response. There has been no additional attempt to effectively bridge the gap between scientific enquiry and human welfare gain at the end-user level. It gives a signal that at the governance level there is lack of awareness that no behavioural response for new technology is automatic. It has to be induced through policy, economic incentive design and designing appropriate institutions. Institutions need to be understood in true sense of the term as representation of rules, regulations and options. An engineering solution-dominated approach remained relatively weak in science of social dynamics and role of behavioural response completely ignored the demand-side management issues. A similarity can be drawn with example of a hardware system with no software loaded. Hardware of water supply systems has been put in place but in the absence of any economic incentive design and policy framework, the system lacks appropriate software and has led to non-functionality of the hardware and gradual depletion. After couple of decades effort technology deployment motive and technology adoption are still apart.

Any future action can start from the past lesson and with a goal of sustenance of projects beyond project lifetime. This calls for appropriate policy planning which goes beyond mere technology development and deployment and should start even before technology choice backed by rigorous goal-oriented socio-economic survey and behavioural analysis. Major fault with the approach lies in its top-down approach and too technocratic view which alienates one from the all-important social goal. The step-by-step approach remained disproportionately small in scale of actions relative to the magnitude of the problem. So it could not generate required momentum for take-off towards sustainability.

Comprehending Costs and Benefits of Arsenic Mitigation Measures: Case Study of Murshidabad District

To assess cost-effectiveness and benefit–cost ratios we have focused on one district. As data need for such assessment is high we could not do for all districts.

In Murshidabad district, over last one decade at various time intervals, a total of Rs. 4744.87 lakhs (4136.87 lakhs for ring well, ATUs and PWSS; 608 lakhs for tube well installation) has been invested for installation of various mitigation technologies (4136.87 lakhs for ring well, ATUs and PWSS; 608 lakhs for tube well installation) with expected beneficiaries totalling 18.87 lakh people (or 3.78 lakhs households). Table A.15 shows mitigation technologies with associated installation one-time costs and total population projected to derive benefit in Murshidabad district. The number of beneficiaries is calculated based on the calculations of any projects of PHED uses as rule of thumb (i.e. 250 persons/tube well (with or without ATU)/ dug well and for PWSS actual beneficiaries as reported in the PHED, Murshidabad are considered). We calculate the number of households benefited using average family size of five based on field study. It needs to mention that household may be using more than one source. Simple sum will lead to overestimate of beneficiaries. This is observable from the statistics also as total beneficiaries workout to 7 million approximately as compared to total population of 5 million plus. Field observations could also show that while some areas are over served by various technologies some areas are deprived of the benefits of installations. To avoid double counting we assume actual beneficiaries are 10 % less than total aggregate sum.

Plausible assumption based on our schematic diagram (Fig. A.1) is that blocks with higher average arsenic concentration level must have higher arsenic risk so per capita investment should be as per risk level. But we find the value of correlation coefficient between average arsenic concentration and per capita investment in arsenic-affected blocks of Murshidabad district is very low ($r = 0.025$) implying that actual implementation and investment have not (Table A.16) followed the water quality information.

We use this number and lifetime of the technologies to arrive at the annualised installation cost of the individual technologies (Table A.17) to compare them across various technologies. If we compare just the annualised costs, then we will rank dug well, tube well, ATU and piped water supply technologies in ascending order of cost. But mere accounting cost consideration will depict a misleading picture if beneficiaries projected to be served by each are not considered. Under ideal conditions (total projected beneficiaries served each day, technology provide year round service), we estimate the annualised capital cost to provide arsenic safe water per beneficiary household (Table A.17) and find that ranking of technologies by least-cost criterion changes and the sequence in ascending order is tube well, dug well, ATU and PWSS. However, we need to understand that just cost-effectiveness

Table A.15 Total costs of different mitigation measures (except tube well installation) and the projected number of beneficiaries

Sl. No.	Name of the block	Different Mitigation Measures						Total Installation Costs (Lakh)	Total population benefitted
		Ring well		ATU		PWSS			
		Total Installation Costs (Lakh)	Population Benefitted	Total Installation Costs (Lakh)	Population Benefitted	Total Installation Costs (Lakh)	Population Benefitted		
1	Nowda	3.85	1750	24.5	8750	60.20	84086	88.55	94586
2	Hariharpara	4.4	2000	37.8	13500	76.95	71567	119.15	87067
3	Beldanga-I	23.65	10750	129.5	46250	101.89	48750	255.04	105750
4	Berhampur	3.85	1750	109.2	39000	190.97	45563	304.02	86313
5	Beldanga-II	0	0	14	6000	141.58	26650	155.58	32650
6	Farakka	0.55	250	8.4	3000	32.59	30528	41.54	33778
7	Suti-I	1.65	750	23.1	8250	184.26	28496	209.01	37496
8	Suti-II	1.65	750	18.9	6750	78.48	34997	99.03	42497
9	Samserganj	0	0	8.4	3000	60.52	48971	68.92	51971
10	Raghumathganj-I	0	0	0	0	90.86	17611	90.86	17611
11	Raghumathganj-II	1.65	750	28	10000	223.63	27233	253.28	37983
12	Bhagwangola-I	4.95	2250	46.9	16750	259.46	27154	311.31	46154
13	Bhagwangola-II	4.95	2250	32.9	11750	43.60	27835	81.45	41835
14	M-J Block	1.65	750	32.9	11750	8.66	5576	43.21	18076
15	Lalgola	0	0	23.1	8250	18.38	38267	41.48	46517
16	Jalangi	2.75	1250	69.3	24750	1019.48	159409	1091.53	185409
17	Raninagar-I	1.65	1250	43.4	15500	87.11	46589	132.16	63339
18	Raninagar-II	2.75	750	62.3	22250	337.18	107473	402.23	130473
19	Domkal	2.75	1250	74.90	26750	270.87	91775	348.52	119775
TOTAL		**62.7**	**28500**	**787.5**	**282250**	**3286.67**	**968530**	**4136.87**	**1279280**

Source PHED (Murshidabad) Report (2003, 2007)

Table A.16 Average arsenic concentration and per capita investment in arsenic-affected blocks of Murshidabad District

Name of the blocks in Murshidabad District	Per capita investment (Rs.)[a]	Average arsenic concentration (μg/L)[b]
Nowda	93.62	161
Hariharpara	136.85	211
Beldanga-I	241.17	208
Berhampur	352.23	118
Beldanga-II	476.51	124.5
Farakka	122.98	89
Suti-I	557.42	141
Suti-II	233.03	168
Samserganj	132.61	110
Raghumathganj-I	515.93	137
Raghumathganj-II	666.82	125
Bhagwangola-I	674.50	150
Bhagwangola-II	194.69	195
M-J Block	239.05	117
Lalgola	89.17	130
Jalangi	588.71	244
Raninagar-I	208.66	159
Raninagar-II	308.29	264
Domkal	290.98	191

[a]Calculated on the basis of Table A.15
[b]Author's estimation

or least cost cannot be a good criterion for technology ranking because benefits generated by each are different although each is targeted to provide arsenic safe water. The difference in benefits arises from consumer-friendly nature of the technology, accessibility with ease, water handling flexibility, etc. However, such detailed technology specific analysis is outside the purview of this thesis.

It is important to evaluate the achievement of these programmes in terms of benefit–cost ratio. In the economic valuation of arsenic safe water, Roy (2008) shows that arsenic removal by 1 μg per litre of water can generate benefits at the household level of Rs. 0.49 (on an average) per month. If the arsenic concentration is reduced to the safe limit of 50 μg/L from an average concentration level of 655 μg/L, the benefit to each household is Rs. 297 per month while the annual household gain is Rs. 3564. The same benefits are Rs. 161 per month and Rs. 1934 per year if the arsenic concentration is reduced by half. By benefit transfer principle we can assess economic benefit generated by the arsenic safe water supply programmes as per claim of the PHED. In case of Murshidabad district, if we reduce the arsenic concentration to the safe limit (50 μg/L) from the average concentration (240 μg/L), then the total welfare gain will be Rs. 93 per household per month and annually Rs. 1117 per household. These numbers are lower than

Table A.17 Annualised total costs and benefits from different mitigation measures

Measures	Total number	Life time	Discount rate (%)	Total annualised capital cost[a] (Rs./installation).	Annualised cost per beneficiary household (Rs./household)
Dugwell	114	10	10	5600	0.69
Tubewell	2432	10	10	9800	0.54
ATU	1131	3	10	29,700	2.72
PWSS	47	30	10	23,140,000	8.24

Source Authors' estimates

[a]Formula used:

$P = A[1 - (1 + r)^{-n}/r]$

where A = Annualsed Cost of capital

n = Life span

r = Discount rate

North 24 Parganas (Roy 2008) because of relatively lower average concentration in Murshidabad. The total number of households at risk in the arsenic-affected areas of Murshidabad district is 0.79 million. Total welfare gain from any arsenic removal mission can generate benefit of Rs. 74 million per month if we cover all the risk population by the arsenic safe water. It is important for the policy makers to know that reduction in arsenic concentration in water to the safe limit through technological and policy interventions can generate such huge benefits. These benefits include both direct benefits and indirect health benefits. If we consider the Mitigation actions over the period of 1997–2007 in Murshidabad and consider the projected beneficiaries (Table A.15), then the monetary value of the annual benefit works out to Rs. 3.9 crores generated by total investment of Rs. 146 crores whose annualised value is around 2.3 crores. So the benefit–cost ratio works out to 1.6. Simple back of the envelope calculation shows that in terms of direct and direct benefit generated payback period is one year.

Above analysis shows the actions so far have lagged far behind the need based on scale of the problem. Cost–benefit and cost-effectiveness analyses show that technology deployment has high potential for benefit–cost ratio that can drive investment towards making sustainable access to arsenic safe water. This analysis is conditional on the fact that benefit recovery mechanism is designed at implementation stage. Although all technologies are feasible by cost–benefit assessment practical implementation will be guided by return such investments can fetch to ensure financial sustainability. In the next two chapters, we try to assess if there is enough demand for arsenic safe water in rural areas through intensive analysis based on individual and household (ultimate user/beneficiary) survey. In the process we try to assess what has been the strength and weaknesses of the implementation programmes adopted so far based on user perception.

Technical Appendix: Methodology

The primary objective of the research survey was to assess the socio-economic condition of arsenic-affected rural poor people of West Bengal, the representative state to understand the impact of the groundwater arsenic contamination catastrophe in West Bengal, and in turn in India. Furthermore, we zeroed in on Murshidabad district, one of the severely arsenic-affected districts with well-documented poverty issues. We visited total 60 villages selected on the basis of arsenic concentration levels reported by Public Health Engineering Department, Government of West Bengal (PHED, WB) and School of Environmental Sciences, Jadavpur University, Kolkata (SOES, JU). In-depth interviews and focus group discussions were used to bring out common perceptions about the impact of arsenic on social as well as economic issues. Different types of questionnaire surveys were conducted to estimate the socio-economic burden of arsenic contamination on this poverty stricken society. Impact on different socio-economic strata of arsenic victims was put under scrutiny. The population investigated especially consisted of male bread winner, young male, married female and young female going to be married as well as arsenic-affected families and their interactions with arsenic-unaffected families. The base information on the subject was obtained from 200 sample households from the selected villages during a period of 2007-12. Purposive sampling procedure was applied, i.e. we visited those households where at least one of the family members are suffering from or died due to arsenic poisoning. A total of 306 patients affected by arsenic were found following this procedure from the sample households. In the focus groups, 'interaction discussion' generated 'rich detail of complex experiences and the reasoning behind actions, beliefs, perceptions and attitudes' (Carey 1995). Qualitative data was used to elicit in-depth material about culture, meanings, processes and problems. In the present study they were employed to deduce people's understandings of the impact of arsenic on their social lives. A total of 30 focus group discussions (FGDs), where each FGD consisted of people from two adjacent villages, were undertaken to verify the problem. Each of the groups consisted of sufferers, non-sufferers and people belonging to different occupations. This method was adopted in order to observe the debate and complexity in social concerns. We also considered the media coverage on arsenic and related issues published in some leading newspapers (since1983) in West Bengal.

For understanding how successfully the arsenic remediation technologies are functioning on the ground, we covered about 20 affected villages from the same district over the period of 2008–2011. Each type of technological options was visited physically to check their functioning status and to observe the usage patterns by targeted beneficiaries. We selected and visited those villages which have high arsenic concentration in the district (using reports of both government and non-government organisations) and people are dependent on tube-well water for drinking water purpose for decades. Techniques for data collection included in-depth key informant interviews and focus group discussions. Both men and women were targeted as informants.

To arrive at the household level units we followed several steps. The first step was to identify blocks and villages with habitations that have been reported to have the highest level of arsenic concentration. We ranked the blocks on the basis of arsenic concentration level. Jalangi and Raninagar-II, the chosen blocks, emerged as the most affected blocks. The villages were selected following the same procedure as the blocks. Subsequently, we visited total 200 families and interviewed each family member, individually, using pretested questionnaire.

The selection of household was done through random sampling. We visited all the chosen habitations and identified the shallow tube wells for which concentration levels were reported. We then listed the households in the command area of each water source and randomly selected the number of households that these tube wells catered to. The number of households surveyed in each command area varied depending on the size and number of residents of the command area.

A key aspect of the survey was to elicit arsenic disease-related information through both direct questioning of the households and the knowledge that we have gathered from the preliminary discussions with expert dermatologists in Kolkata. In the study area we covered, extensively, we found households to be exposed to arsenic-awareness campaigns. Many also knew about their diseases because of visits to arsenic clinic. Subsequently, we also categorised, basis information gathered through our surveys, melanosis as low perceived stage of the disease and keratosis and cancer, etc. as high perceived stages of the disease.

Willingness to Pay (WTP)

We estimated demand for quality drinking water, in the absence of explicit market mechanisms, through alternative methods of economic valuation. Individuals are assumed to be able to value changes in water quality (environmental service), despite absence of the market. Conventional definition of WTP is the maximum sum of money that an individual is willing to pay for a certain good rather than do without an increase in the price of the same. It is postulated that if an improvement occurs and the consumer believes he is better off in some way due to this change, then he may be willing to pay money to secure this improvement. This WTP thus determined reflects his economic valuation of the improved service. This measure of value, which has the theoretical underpinning in the neoclassical welfare economics (Freeman 1993), is what we would like to estimate, in order to formulate policy decisions on how to prioritise and allocate public money or design economic incentive to achieve sustainability in the arsenic mitigation exercise.

For those goods which have a market, the valuation is reflected through demand curves. The demand curve reflects consumer preference by plotting the desired amount of a good against various prices for a given income level. For environmental goods like water quality, the method of capturing preferences has to be modified. Application of standard microeconomic theory for measuring changes in individuals' well being was developed for the purpose of interpreting changes in the prices and quantities of goods purchased from the market.

The theory assumes that individuals have well-defined preferences among alternative bundles of goods comprising both market and non-market goods of varying quantities. It is assumed that the agents know their preferences and there is substitutability among alternative goods in a bundle. Substitutability implies that if the individual reduces the consumption of one good, then it is possible to increase the consumption of the quantity of another good in the bundle so as to leave the individual no worse-off because of the change. The trade-off between different goods ensured by the substitutability is an indicator of the values of goods that people place on these goods or services. These value measures based on substitutability can be expressed in terms of willingness to pay (WTP).

We can measure WTP within the theoretical framework of the conventional consumer behaviour theory pertaining to the welfare impact and subsequent valuation of welfare changes due to a change in the quality of environmental goods or services. A variety of techniques and models have been developed for assigning monetary values to gains or losses associated with changes in the quality of environmental amenities (UN 1997). The valuation can be done from either the demand side or the supply side.

On the demand side, consumers insist on superior quality of environmental amenities like clean air, potable water, etc., since these elements enhance utility and lead to higher welfare levels. Thus valuation can be done from either of the two aspects—how much an individual is willing to pay in order to acquire a certain quality of an environmental good or alternatively, and how much an individual is willing to pay in order to avoid a low quality of any environmental amenity. Either way, we obtain a proxy for the value that a consumer places on an environmental good like drinking safe quality water. Demand-side valuation methods fall under two broad categories—physical linkage method and behavioural method. Physical linkage method includes measures like cost-of-illness (COI) approach, human capital approach and statistical value of life. The behavioural method can be classified based on two different characteristics. First, whether the data is obtained from real-life observations or whether it is under a hypothetical setup, and second, whether the method yields monetary value directly or whether the monetary values have to be inferred through some indirect technique. Hence, any method of estimating environmental and resource values can be placed in one of the four following categories:

(i) Direct observed,
(ii) Indirect observed,
(iii) Direct hypothetical, and
(iv) Indirect hypothetical.

The estimation of WTP for a certain environmental resource is contingent upon the valuation technique adopted for the purpose. We use stated preference to derive WTP. Respondents were asked whether they were willing to get arsenic safe water for their family and willing to pay for the Operation and Maintenance (O&M) of those mitigation measures. 100 % respondents put their positive response for getting arsenic safe drinking water for their family. For value elicitation, respondents were asked to choose a value between Rs. 10 and 100 per month. The survey results show that as contribution charge increases, number of contributors decreases and vice versa. About 8.5 % of the respondents were unwilling to pay mainly because they could not afford the cost. We use the results of this survey to derive the demand curve based on WTP values.

We observe that as price goes up demand falls. It is like any standard good. Ability to pay is examined by looking at the household income and expenditure figures. The average WTP towards O&M of the mitigation measures is Rs. 26. In relation to the income and expenditure, low-income (BPL) group households are willing to pay substantially greater proportion of their incomes for availing access to better quality of water. BPL households are willing to pay 1.1 % of their income as against 0.99 % in case of middle income group, and a mere 0.27 % in case of higher income group. The survey results also pointed out that in terms of expenditures also BPL households are willing to pay more than their counterparts.

WTP values can be used to determine the economic sustainability of providing arsenic-free water. These values are useful for projecting the number of household connections and revenue generation willing to avail the provision of supplying arsenic-free water.

Left to market mechanism we find that not all households will go for access (it may be through connection to piped water system, ATU and the like) to safe water source. This highlights two important policy implications. What can be a possible water service charge for safe water quality can be guessed based on this study and second what might be the possible subsidy level for making provision of arsenic safe water based on possible revenue generation estimates? Highest possible revenue generation from 100 willing/connected households is expected to be at Rs. 720, when access charge is Rs. 40 per month per household. But at this rate only 18 % will opt for connections. But it is important to note that after spiking at Rs. 40 per month per household the revenue generation will fall. So if revenue generation for reinvestment is the goal, water access charges cannot be set beyond Rs. 40 per month per household. However, exact rate depends upon the objective preference of the policy makers. But if goal is to provide access to all then differentiated water access charges may be designed to serve large number of people as well as to generate re-investible fund.

Correlation Analysis

Correlation analysis has been done to check the degree of association between (1) the number of arsenic patients in the family and the dwelling type; (2) perceived stage of the disease and education level, monthly income of the respondent; (3) taking medical treatment and the perceived stage of the arsenicosis (like Melanosis, Keratosis, hyper-keratosis, gangrene) and proportion of male/female arsenic patients in the family; and (4) WTP and source of income, family expenditure, education, number of arsenic patients in the family. Higher the monthly income and the education level, the severity of skin lesions gets lower as it is reflected by a significantly negative value of the correlation coefficient. There exists a positive and significant relation between the perceived stage of the disease and the possibility of opting for medical treatment. During the field survey, we have realised that arsenic patients are going to take proper medical treatment when the skin lesion reaches the stage of keratosis, hyper-keratosis and gangrene. The value of correlation coefficient between household participation in taking medical treatment and the proportion of male arsenic patients in the family is higher than the value of correlation coefficient between household participation in taking medical treatment and the proportion of female arsenic patients in the family. The correlation between WTP and source of income, family expenditure and education is found to be significantly positive. This means that the higher family income, expenditure and education lead to greater contribution towards WTP. The value of the correlation coefficient between WTP and number of arsenic patients in the family is negative (although not significant). This might be a reflection of the fact that arsenic patients are found mostly in poor families whose ability to pay is very low.

Regression Analysis

We have carried regression analysis to discern causal relation and relative strength of determining variables to help in deriving policy implications. We want to see

(a) Whether a family will adopt medical treatment or not. What determines this behaviour?
(b) Whether an individual suffering from arsenicosis is taking medical treatment or not? What determines this behaviour?
(c) Whether a household will be willing to pay for arsenic safe water service received or not? What determines this behaviour?
(d) Determining variables for medical expenditure.

Because of the dichotomous nature of the dependent variable for a, b and c above like whether taking any medical treatment (MT), Willingness to Pay (WTP), a non-linear probabilistic model has been used for estimation. The most widely used model is based on logistic regression analysis. The model helps in estimating the probability of occurrence of an event and is given by

$$P_{\mathrm{i}} = \text{Probability}(\text{event}) = 1/(1 + e^{-z_i})$$

where Z_i is the linear combination of variables $X_1, X_2, X_3, \ldots, X_q$

$$Z_i = \alpha_0 + \alpha_1 X_1 + \alpha_2 X_2 + \alpha_3 X_3 + \cdots + \alpha_q X_q$$

If P_i is the probability of occurrence of the event, then $(1 - P_i)$ is the probability of the event not occurring.

Therefore, we can write

$$e = y^* / y$$

Now, $P_i/(1 - P_i)$ is simply the odd ratio in favour of the event.

The above probability expression can be transformed to determine the log odds in favour of the event as

$$\begin{aligned} L &= \text{Log }[\text{Prob(event)}/\{1 - \text{Prob (event)}\}] \\ &= \alpha_0 + \alpha_1 X_1 + \alpha_2 X_2 + \alpha_3 X_3 + \cdots + \alpha_q X_\text{q} \end{aligned} \quad \text{(A.2)}$$

the log of odd ratio is not only linear in X, but also linear in parameters αi, the slope measures the change in L for a unit change in X, that is, it tells how the log odds in favour of the event.

However, for case (d) above dependent variable for medical expenditure is a continuous variable so have to set up a standard regression model compatible with ordinary least square (OLS) estimation method.

The multiple regression model is given as

$$Y = \beta_0 + \beta_1 X_1 + \beta_2 X_2 + \beta_3 X_3 + \cdots + \beta_q X_\text{q} + U \quad \text{(A.3)}$$

where β is the regression coefficient, Y is the dependent variable, X is the independent variable and U is the error term with BLUE properties. All these regression model estimations have been done using application package STATA (version 10).

Model A: Medical Treatment (MT) Seeking Behaviour

Based on our field survey and correlation analysis, we postulate that whether a family participates in medical treatment for their arsenic-affected patients or not that depends on

(i) Level of family education as a proxy of awareness (TFEDU),
(ii) Type of Dwelling as a proxy of income level of household (DWT),
(iii) Proportion of arsenic patients in the family (PROASP), and
(iv) Perceived stage of arsenicosis (PSTAAS).

$$Z = \alpha_0 + \alpha_1(\text{TFEDU}) + \alpha_2(\text{DWT}) + \alpha_3(\text{PROASP}) + \alpha_4(\text{PSTAAS}) \quad (A.4)$$

Estimated values of the parameters α_1 through α_4 can be used to describe the probability of participating in the medical treatment by the family for their arsenic patients. In our analysis, we consider education as a proxy variable of awareness level because of the general perception that level of education positively affects the awareness level. Total family education vis-à-vis awareness of the family positively affects decision towards availing medical treatment. Income is another important factor influencing the crucial decision of participating in the medical treatment. But in many cases people do not reveal their actual family income, so type of dwelling is considered as a proxy variable of income in our estimation. Our field-level experience suggests that the higher the level of family education and the better the dwelling type, the higher is the chance of their participation in the medical treatment. In the course of our field survey, we have observed that families with higher proportion of arsenic patients have greater chance to participate in the medical treatment. Not only that most of the surveyed families decided to get medical treatment only when patients' health conditions reach the extreme stage (like keratosis and gangrene).Here we consider this variable as dummy variable. It takes value '1' for participating in the later stage of health condition and '0' for participating in the early stage of health condition.

Empirical analysis through the estimation of logit model helps us to get the nature and magnitude of the coefficients of Eq. (A.4).

Table A.18 gives the estimated coefficient values of the variables and corresponding *p* values. The variable TFEDU (education level) has expected sign and statistical significance. The probability of household to be participated or not in the medical treatment depends on education vis-à-vis awareness level of the household established. Coefficient of dwelling type (proxy of income level) appears with correct sign but is statistically insignificant. This may be due to the fact that low-income-level families often participate in medical treatment by taking loans or by selling their physical assets (like trees, animals, etc). The variable proportion of arsenic patients in the family (PROASP) and perceived stage of arsenicosis (PSTAAS) has expected sign and statistical significance. The probability of household to be participated in the medical treatment depends on the proportion of the arsenic patients in the family and perceived stage of the arsenic patient in the family is established. As our model assumes a non-linear functional relationship between the dependent variable and independent variables, we report (Table A.19)

Table A.18 Parameter estimates

Variable	Coefficient	*p* value
TFEDU	0.0313326	0.018
DWT	−0.0001108	0.214
PROASP	1.199603	0.053
PSTAAS	1.718091	0.000

Table A.19 Marginal effects

Variable	Coefficient	p value
TFEDU	0.0078066	0.018
DWT	−0.0000276	0.214
PROASP	0.2988837	0.053
PSTAAS	0.3962196	0.000

the marginal effects of each of the independent variables. These marginal effects can be interpreted as the increment in the probability of taking medical treatment due to an increase in any one of the independent variables by 1 %. The statistical results of Table A.19 explain that except for one variable (DWT) other variables have a significant marginal contribution on the dependent variable.

Model B: Taking medical Treatment (MT): Individual Level Analysis

In the present context, it is postulated that whether a family member participating in medical treatment or not depends on

(i) Monthly income (MI),
(ii) Perceived stage of arsenicosis of the individual (PSTAASI),
(iii) Gender of the individual (GENDER).

Consider the following model:

$$Z = \alpha_0 + \alpha_1(\text{MI}) + \alpha_2(\text{PSTAASI}) + \alpha_3(\text{GENDER}) \tag{A.5}$$

Here MI is quantitative variable (an important factor of participating in the medical treatment. It is expected that the higher the level of income, the higher would be the chance to take medical treatment) and PSTAASI, GENDER are qualitative or dummy variables. Qualitative variables are defined as

$$\begin{aligned}\text{PSTAASI} &= 1 \text{ for high/critical stage}\\ &= 0 \text{ for low/initial stage}\\ \text{GENDER} &= 1 \text{ for male}\\ &= 0 \text{ for female}\end{aligned}$$

Estimated values of the parameters α_1 through α_3 can be used to describe the probability of participating in the medical treatment by an individual for arsenic treatment. Our field-level experience revealed that an individual takes medical treatment only at the high/critical stage of the disease. Again there exists a clear gender bias towards taking medical treatment, i.e. males participate more in taking medical treatment among the total number of patients taking medical treatment. Empirical analysis through the estimation of logit model helps us to get the nature and magnitude of the coefficients of Eq. (A.5).

Table A.20 gives the estimated coefficient values of the variables and corresponding *p* values. The variable monthly income (MI) has expected sign but statistically insignificant. The variable perceived stage of arsenicosis (PSTAAS) and GENDER has expected sign and statistically significant. The probability of an individual to be participated in the medical treatment depends on the perceived stage of the arsenic disease of the patient as well as on the gender is established. Table A.21 shows the estimated marginal effects of each of the independent variables. These marginal effects can be interpreted as the increment in the probability of taking medical treatment due to an increase in any one of the independent variables by 1 %. The statistical results of Table A.21 explain that except for one variable (MI) other variables have a significant marginal contribution on the dependent variable.

Our field-level experience also revealed that the probability of taking medical treatment is high among the male patients who were suffering from high stage of arsenicosis. The same is true for female patients also. For this we define four mutually exclusive categories: male patients with high stage of arsenicosis, male patient with low stage of arsenicosis, female patient with high stage of arsenicosis and female patient with low stage of arsenicosis. To capture this we have introduced three dummies. We can use the following model:

$$Z = \alpha_0 + \alpha_1 D_1 + \alpha_2 D_2 + \alpha_3 D_3 \tag{A.6}$$

where

Table A.20 Parameter estimates

Variable	Coefficient	*p* value
MI	8.27e−06	0.931
PSTAASI	0.931	0.000
GENDER	1.103411	0.003

Table A.21 Marginal effects

Variable	Coefficient	*p* value
MI	1.98e−06	0.931
PSTAASI	0.4236201	0.000
GENDER	0.2422955	0.001

Table A.22 Parameter estimates

Variable	Coefficient	p value
D_1	2.978925	0.000
D_2	1.235063	0.002
D_3	2.184802	0.001

Table A.23 Marginal effects

Variable	Coefficient	p value
D_1	0.6249322	0.000
D_2	0.2881047	0.001
D_3	0.47066	0.000

$$
\begin{aligned}
D_1 &= 1 \text{ for male patients with high stage of arsenicosis} \\
&= 0 \text{ otherwise} \\
D_2 &= 1 \text{ male patient with low stage of arsenicosis} \\
&= 0 \text{ otherwise} \\
D_3 &= 1 \text{ female patient with high stage of arsenicosis} \\
&= 0 \text{ otherwise}
\end{aligned}
$$

Here we are arbitrarily treating the "female patient with low stage of arsenicosis" category as the base category. Therefore, the intercept α_0 will reflect the intercept for this category. The differential intercepts α_1 and α_2 tell by how much the intercepts of the two categories differ from the intercept of the base category (Tables A.22 and A.23).

Model C: Willingness to Pay (WTP): Household Level Analysis

WTP is the valuation placed by an individual on a good or service in terms of money. Here we first calculated the median WTP and then we consider this variable as dummy variable. It takes value '1' for households willing to pay more than median WTP and '0' for households willing to pay less than median WTP. Now we run a binary regression using median WTP (Z) as dummy dependent variable and total family income (TFI), total family education (TFEDU), per capita expenditure for arsenic treatment (PCEASTREAT) and proportion of arsenic patients in the family (PROASP) as independent variables.

$$Z = \alpha_0 + \alpha_1(\text{TFI}) + \alpha_2(\text{TFEDU}) + \alpha_3(\text{PCEASTREAT}) + \alpha_4(\text{PROASP}) \quad (A.7)$$

Estimated values of the parameters α_1 through α_4 can be used to describe the probability of WTP for arsenic safe water.

It is postulated that higher WTP is associated with higher level of education and higher level of family education. Again proportion of arsenic patients with high perceived level of arsenicosis and per capita expenditure for the treatment of arsenic patient is higher in low-income families. Due to low capacity to pay, it is expected that families with higher proportion of arsenic patients and higher per capita expenditures for arsenic treatment have low WTP.

Empirical analysis through the estimation of logit model helps us to get the nature and magnitude of the coefficients of Eq. (A.7).

Table A.24 gives the estimated coefficient values of the variables and corresponding p values. The variable TFI has expected sign but statistically insignificant. The probability of household to be paid more than median WTP is high for high-income group and low for low-income families. Coefficient of total family education (TFEDU) appears with correct sign and statistically significant. The variable proportion of arsenic patients in the family (PROASP) and per capita expenditure for arsenic treatment (PCEASTREAT) has expected sign and statistical insignificance.

Table A.25 indicates the estimated marginal effects of each of the independent variables. These marginal effects can be interpreted as the increment in the probability of paying more money towards arsenic safe water due to an increase in any one of the independent variables by 1 %. The statistical results of Table A.25 explain that except for one variable (TFEDU) other variables have an insignificant marginal contribution on the dependent variable.

Table A.24 Parameter estimates

Variable	Coefficient	*p* value
TFI	0.000213	0.070
TFEDU	0.0560691	0.000
PCEASTREAT	−0.000186	0.199
PROASP	−1.004646	0.121

Table A.25 Marginal effects

Variable	Coefficient	*p* value
TFI	0.000053	0.069
TFEDU	0.0139665	0.000
PCEASTREAT	−0.0000463	0.199
PROASP	−0.2502518	0.121

Model D: Per Capita Medical Expenditure: Household Level Analysis

We postulate that the per capita expenditure for arsenic treatment is highly related to the families where patients have high perceived stage of the arsenicosis. Here we consider perceived stage of the arsenicosis as dummy variable. It takes value '1' for participating in the later stage of health condition and '0' for participating in the early stage of health condition, and it is high for the low-income families. It is also postulated that the high medical expenditure has strong gender bias, which is mainly for male affected persons.

To see this we have run a simple regression by considering per capita medical expenditure as dependent variable and the following as independent variables:

(i) Total family income (TFI),
(ii) Perceived stage of arsenicosis (PSTAAS),
(iii) Proportion of male patient taking treatment (PMPT).

$$Z = \alpha_0 + \alpha_1(\text{TFI}) + \alpha_2(\text{PSTAAS}) + \alpha_3(\text{PMPT}) \quad \text{(A.8)}$$

Estimated values of the parameters α_1 through α_3 can be used to describe the slope coefficient of a regressor which measures the effect on the average value of the regressand for a unit change in the value of the regressor. Empirical analysis through the estimation of regression model helps us to get the nature and magnitude of the coefficients of Eq. (A.8).

Table A.26 gives the estimated coefficient values of the variables and corresponding p values. The variable total family income (TFI) has expected sign but not statistically significant. The variable perceived stage of arsenic patients in the family (PASTAAS) and PMPT (Proportion of male patients taking medical treatment) has expected sign with statistical significance.

It is clear from individual as well as household level study that there is enough awareness about need for access to arsenic safe water and people are willing to pay for the service. It is now an important policy decision how a safe water access charge may be designed that can generate revenue for sustaining service delivery as well as to provide access to safe water system to all.

Table A.26 Parameter estimates

Variable	Coefficient	*p* value
TFI	−0.0210435	0.620
PSTAAS	588.0974	0.002
PMPT	595.3748	0.002

Survey Questionnaire

Name of the Investigator	
Date of Survey	

1. Area Details

District :	
Block :	
G.P/ Thana	
Village	

2. Information on Respondent

Name of the respondent	
Age	
Sex	
Education	
Is the respondent head of the family?	
If no, then what is the relationship with household head?	
Occupation	
Income (Daily and Monthly)	

3. Information on Individual Household Member

SL. No.	Name of the Member	Sex	Age	Educational attainment	Other Skill (Diploma, training etc.)	Presently engaged in	
						Education	Occupation[1]
1.							
2.							
3.							
4.							
5.							
6.							
7.							
8.							
9.							
10.							

(1) (a) Agriculture(Cultivator/Agricultural Labour/Both) (b) Daily Labour (c) Business (d) Self-employed (e) Service

4. General Information on Household

Name of the head of the family		
Age		
Education		
Occupation		
Number of family members		
Number of income earners in the family		
Main source of family income		
Type of Dwelling	Floor………………, Roof…………………	Wall……………………
Having electric connection (Y/ N)		
Electricity bill (if any)		
Type of Latrine you use	Sanitary …… Not sanitary………Open Defecation……………………………	
Any family member stay outside for job (How many?)		
If 'Yes' then sent money for family expenditure		
If 'Yes' then amount	Monthly	Yearly
Total family income	Daily	Monthly
Total family Expenditure	Daily	Monthly
Floor area		
Fuel consumption detail		

5. Food Habits of the Household

Items	Amounts		
	Day	Week	Month
Rice			
Potato			
Onion			
Dal			
Vegetables			
Cooking Oil			
Fish			
Egg			
Meat			
Milk			
Fruits			

6. Respondents awareness of the Arsenic problem and Related Questions

Is there any Arsenic Problem in your Village? (Y/N)		
When do you Know (mention year)?		
Source of Information?	1. Newspaper ....2.TV............3.Radio............4. Residents of the village......... 5. Relatives.......... 6. Govt.........8. NGO.........9.Others.............................	
Do you consider Arsenic as a problem?(Y/ N)		
If Yes, Why?		
If No, Why?		
Are there any arsenic patients in your village?		
Was there any awareness programme held for arsenic in the neighbourhood? (Y/N)		
If Yes, then mention how many such programmes held in your locality (Also mention Year)?		
Who organized such programme?	NGO ( ), Govt ( ), GP ( ), Others ( )	

7. Testing Demand for Arsenic

Have you any privately owned tubewell? (Y/ N)	
Have you tested your tubewell for Arsenic?(Y/N)	
If No, then why?	
If Yes, then from Where (Also mention Year)?	
Do you have to pay money for testing? If Yes, how much (in Rs.)?	
What was the test result?	

8. Water Supply-Related Questions

What was your source of drinking/ cooking water before knowing the Arsenic problem?		Pvt. TW ( ), Govt. TW ( ), Pond ( ), Dugwell ( ), PWSS ( ), Others ( )		
How many years are you using that source for drinking/ cooking water?				
Are there any alternative options installed to supply Arsenic free water in your locality?(Y/N)				
If Yes, then fill-up the following table:				
Options	When install	Are they in running condition	How long they are running well	Are you collecting water from these sources
ATU				
Deep Tubewell				
Dugwell				
Swajaldhara				
PWSS				
What is your present source of drinking/ cooking water?				
Distance to source (in Km)?				
Daily time spent to collect water (in hours/ minutes)				
Who collects water?		Family members	Hired	
Do you have to pay money to collect water? If, yes, then amount (in Rs.)		Daily	Monthly	
If hired, Procurement Cost (Bhari) (in Rs)		Monthly	Annually	
Purchased safe water is used for what purpose		Only Drinking	Only Cooking	Both
Purchased safe water is used by		All family members	Selected family members	
Is there any privately owned alternative option (like Domestic Filter) in your family? (Y/ N)				
If Yes, then Fill-up the following table				

Option	Year of purchase	From where you purchase	Purchasing cost	Are you using properly	If not, Why?

9. Water Demand-Related Questions

Quantity of water used per day by household (In Litres)?	General	Dinking	Cooking
Are you satisfied with the present source? (Y/ N)			
Reasons for satisfaction/not satisfaction			
Preference Ranking of the alternative mitigation options to collect arsenic safe water	Options	Rank	
	(a) ATU (b) Dugwell (c) Govt. Tubewell (d) Swajaldhara (e) PWSS		
Mention the reasons for such preference:			

10. Price-Related Questions

How much money you are ready to pay for the survival of your preferred mitigation measures?	Minimum (per month)	Maximum (per month)
Which alternative approach of the cost recovery option for the sustainability of water supply scheme you support		
(a) Recovery of O & M cost only		
(b) Recovery of O & M cost Plus interest on Capital Invested		
(c) Recovery of O & M cost Plus interest on Capital Invested Plus Depreciation		
(d) None		

11. Arsenic-Related Medical Expenditures for the Household

Are there any arsenic patients in your family? If Yes, then how many?	
When you detect arsenic patients in your family (mention year)?	
Is their any medical treatment for the arsenic patients in your family?	
If Not treated, then why?	
If yes, then from where?	
Are they satisfied with treatment facilities?(Y/N)	
How you manage money for treatment expenditure?	
Have you sold any assets to treat arsenic patients in your family?	
Are you taking any loan to tackle the expenditures for arsenic treatment in your family?	

12. Arsenic-Related Medical Expenditures for the Household Members

Sl. No.	Name of the arsenic Patients	Age	Sex	When detected (Year)	Who detected	Taking any Medical Treatment	If Yes, then where
1.							
2.							
3.							
4.							

Sl. No	Expenditure on Treatment			Still continuing the treatment (Y/ N)	If not, why
	Doctor's Fee	Medicine Cost	Transport Cost		
1.					
2.					
3.					
4.					

13. Sickness Details of the Household Members

Sl. No.	Name of the patients	Number of days sick spent sick	Length of use of drinking water from same source before attack(in years)	Length of use of drinking water from same source after attack(in years)	Number of Working days lost	Income lost (Monthly/ Annually)
1.						
2.						
3.						
4.						

14. Assets Owned by Households

Assets	Yes/ No	If 'Yes', then specify Number	Is it your source of income? (Y/N)	If 'Yes', then specify income earned from the source for the last year.
Bicycle				
Motor Cycle				
Van Rickshaw				
Telephone				
Mobile Phone				
Television				
Radio				
Kerosene Stove				
LPG Gas Stove				
Goat				
Cow				
Pond				
Land for Cultivation				
Others (Mention)				

References

Evaluation Study On Rajiv Gandhi National Drinking Water Mission (RGNDWM) by Programme Evaluation Organisation, Planning Commission, Government of India, New Delhi, November, 2010. http://planningcommission.nic.in/reports/peoreport/peoevalu/peo_rgndwm.pdf. Accessed 16 Aug 2014

Adak SK, Mandal BK (1999) Yield of potato as influenced by arsenic contaminated irrigation water. In: Abstracts of the global conference on potato, New Delhi, 6–11 Dec 1999, p 149. http://users.physics.harvard.edu/~wilson/arsenic/conferences/soilcrop.doc. Accessed 12 June 2014

Adak SK, Mandal BK (2000) Influence of arsenic contaminated irrigation water on yield *vis-à-vis* uptake in different plant parts. In: Proceedings of the international conference on Managing natural resources for sustainable agricultural production in the 21st century (Extended Summaries, 2: 537), New Delhi, 14-18 Feb 2000

Ahmed MF (eds) (2003) Arsenic contamination: Bangladesh perspective. Published by ITN Bangladesh, Centre for Water Supply and Waste Management, BUET, Dhaka, Bangladesh, pp 481–509

Ahmad SA, Sayed MHS, Khan MH, Karim MN, Haque MA, Bhuiyan MSA, Rahman MS, Faruquee MH (2007) Sociocultural aspects of arsenicosis in Bangladesh: community perspective. J Environ Sci Health Part A 42(12):1945–1958

Backward Regions Grant Fund Programme Guidelines: Prepared and Circulated by Ministry of Panchayati Raj, Government of India, Release Date: 15 Jan 2007. Available at http://www.nird.org.in/brgf/doc/BRGFFINALGUIDELINES.pdf. Accessed on 15 Aug 2015

Barkat A (2004) Socio-economic consequences of arsenic poisoning in drinking water: an issue of unprecedented cultural emergency in Bangladesh. Presented at 32nd session of the international seminars on planetary emergencies, Ettore Majorana International Foundation & Centre for Scientific Culture, in collaboration with World Federation of Scientists & ICSC-World Laboratory

BGS DPHE (2001) Arsenic contamination of groundwater. In: Kinniburgh DG, Smedley PL (eds) Bangladesh, summary. British Geological Survey WC/00/19. British Geological Survey, Keyworth, UK, pp 1–15

Blagbrough V (ed) (2001) Looking back: the long-term impacts of water and sanitation projects. Water Aid, London

Brinkel J, Mobarak MD et al (2009) A Systematic review of arsenic exposure and its social and mental health effects with special reference to Bangladesh. Int J Environ Res Public Health 6:1609–1619

Chakraborty D et al (2004) Groundwater arsenic contamination and its health effects in the Ganga-Meghna-Brahmaputra plain. J Environ Monit 6:75–83

Chakraborti D, Das B, Rahman MM et al (2009) Status of groundwater arsenic contamination in the state of West Bengal, India: a 20-year study report. Mol Nutr Food Res 53:542–551

A. Das et al., *Socio-Economic Analysis of Arsenic Contamination of Groundwater in West Bengal*, India Studies in Business and Economics,
DOI 10.1007/978-981-10-0682-1

Chakraborti D, Das B, Murrill MT (2011) Examining India's groundwater quality management. Environ Sci Technol 45:27–33

Chandrasekharam D (2007) Pathway of arsenic from water to food, West Bengal, India. In: Aswathanarayana U (ed) Food and water security: an integrated strategy for food security in the developing countries. Taylor and Francis, London, pp 63–70

Carey MA (1995) Concerns in the analysis of focus group data : comment. Qual Health Res 5:487–495

Census (2001) http://www.censusindia.gov.in/Census_Data_2001/Census_Data_Online/Social_and_cultural/Religion.aspx. Accessed on 12 Feb 2013

Census (2011) http://censusindia.gov.in/2011-prov-results/indiaatglance.html

Das A (2006) Arsenic pollution- source, problem and solution, Yojana (Bengali Version, ISSN 0971-8435), May, pp 41–45

Das A (2009) The use of arsenic removal plants to mitigate the arsenic in West Bengal (India): a community perspective. Final report of minor research project (UGC sponsored, MRP No.F. PHW-099/06-07 Dated 15th Jan 2007) submitted to the UGC

Das A (2013) Socio-economic accounting of arsenicosis in a gram panchayet (cluster of villages) in Murshidabad District, one of the nine arsenic affected districts in West Bengal. Final report of minor research project (UGC sponsored, MRP No.F.PHW-119/11-12 Dated 8thAug 2011) submitted to the UGC

Das A (2012) The economic analysis of arsenic in water: a case study of West Bengal, Ph.D Thesis submitted at Jadavpur University.

Das A, Das B, Mukherjee SC et al (2013) Groundwater: arsenic contamination. Encyclopedia of Environmental Management, Taylor and Francis, New York. Published online: 01 May 2013, pp 1262–1280 (to link to this chapter: doi:10.1081/E-EEM-1200466645)

Das A, Chakraborti S (2013) The price of life. Int J Human Social Sci Invention 2(2):22–27

Das A, Roy J (2013) Socio-economic fallout of arsenicosis in West Bengal: a case study in Murshidabad District. J Ind Soc Agri Stat 67(2):273–285

Das A, Roy J (2015) Jale arsenic- prasanga paschimbangya. Mitram, College Street, Kolkata

Dangleben NL et al (2013) Arsenic immunotoxicity: a review. Environ Health 12(1):73

Doctor's Handbook (2005) Flurosis-an easily preventable disease through practice of interventions, contribution presented by Flurosis Research and Rural Development Foundation, New Delhi

Dutta DV (1976) Arsenic and non-cirrhotic portal hypertension (letter).Lancet 433

EPA 815-R-00-026 (2000) Office of Ground Water and Drinking Water: US Environmental Protection Agency, Washington D C: Arsenic in Drinking Water Rule: Economic Analysis. http://www.epa.gov/safewater/arsenic.html.

Freeman AM 111 (1993) The measurement of environmental and resource values: theory and methods, resource for the future, Washington, DC

Flanagan Sara V et al (2012) Arsenic in tube well water in Bangladesh: health and economic impacts and implications for arsenic mitigation. Bull the World Health Organ 90:839–846. doi:10.2471/BLT.11.101253. (http://www.who.int/bulletin/volumes/90/11/11-101253/en/). Accessed on 13 Oct 15

Gadgil A, Roy J, Addy S et al (2012) Addressing arsenic poisoning in South Asia. Solutions 3 (5):40–45

García-Esquinas E et al (2013) Arsenic exposure and cancer mortality in a US-based prospective cohort: the strong heart study. Cancer Epidemiol Biomarkers Prev 22(11):1944–1953

Ghosh G, Majumdar K et al (1995) Water supply in rural India (policy and programme). Ashish Publishing House, New Delhi

Gilbert-Diamond D et al (2013) A population-based case–control study of urinary arsenic species and squamous cell carcinoma in New Hampshire, USA. Environ Health Perspect 121 (10):1154–1160

Chakraborti Garai R AK et al (1984) Chronic arsenic poisoning from tube-well water. J Ind Med Assoc 82:34–35

Guha Mazumdar DN, Chakraborty AK, Ghose A et al (1988) Chronic arsenic toxicity from drinking tube-well water in rural West Bengal. Bull WHO 66(4):499–506

Guha Mazumder DN, Haque R, Ghosh N et al (1998) Arsenic levels in drinking water and the prevalence of skin lesions in West Bengal, India. Int J Epidemiol 27:871–877

Guha Mazumder DN, Ghoshal UC, Saha J et al (1998) Randomized placesbo-controlled trial of 2,3 dimercaptosuccinic acid in therapy of chronic arsenicosis due to drinking arsenic contaminated subsoil Water. J Toxicot Clin Toxicol 36(7):683–690

Guha Mazumder, DN, De BK, Santra A et al (1998c) In clinical manifestations of chronic arsenic toxicity. Its natural history and therapy: experience of study in West Bengal, India. In: Proceeding of the international conference on arsenic pollution of groundwater in Bangladesh: causes, effects and remedies, Dhaka Bangladesh, pp 91-93, 8–12 Feb 1998

Guha Mazumder DN, De BK, Ghoshal UC et al (1999) Chronic arsenic toxicity: epidemiology, natural history and treatment. In: Chappell WR, Abernathy CO, Calderon RL (eds) Arsenic exposure and health effects. Elsevier, New York, pp 335–347

Guha Mazumder DN, Haque R, Ghosh N et al (2000) Arsenic in drinking water and the prevalence of respiratory effects in West Bengal, India. Int J Epidemiol 29(6):1042–1047

Guha Mazumdar DN, Sarkar S (eds) (2012) Arsenic contamination in water and food chain. DNGM Research Foundation, Kolkata

Hadi A and Parveen, R (2004) Arsenicosis in Bangladesh: prevalence and socio-economic correlates. Public Health 118(8): 559–564

Hanchett S (2004) Social aspects of arsenic contamination in drinking water: a review of knowledge and practice in Bangladesh and West Bengal. Report for the arsenic policy support unit, local government division, Government of Bangladesh

Hassan MM, Jand Atkins P, Dunn CE (2005) Social implications of arsenic poisoning in Bangladesh. Soc Sci Med 61(10):2201–2211

Health on the March (2001-2002, 2005-2006, 2006–2007) State Bureau of Health Intelligence, Directorate of Health Services, Government of West Bengal, Swasthya Bhawan, Salt Lake, Kolkata

Indian Standards for Drinking Water, second revision of IS 10500, (2004) Available at http://bis.org.in/sf/fad/FAD25(2047)C.pdf. Accessed on 15 Aug 15

International Institute for Population Sciences, IIPS (2010) District level household and facility survey (DLHS-3), 2007-08: India Mumbai (available on http://www.rchiips.org/pdf/india_report_dlhs-3.pdf. Accessed 16 September 2014

Intergovernmental Panel on Climate Change IPCC (2000) Climate Change—Impacts, adaptations and mitigation of climate change: Scientific–Technical Analysis, Cambridge. https://www.ipcc.ch/publications_and_data/publications_and_data_reports.shtml. Accessed 16 Sept 2014

Khan MZH (2007) Managing the arsenic disaster in water supply: risk measurement, costs of illness and policy choices for Bangladesh, South Asian Network for Development and Environmental Economics (SANDEE) Working Paper No. 27-07. Available on website www.sandeeonline.org)

Maharjan M et al (2007) Mutual interaction between nutritional status and chronic arsenic toxicity due to groundwater contamination in an area of Terai, lowland Nepal. J Epidemiol Community Health 61(5):389–394

Mahmood SAI, Halder AK (2011) The socioeconomic impact of arsenic poisoning in Bangladesh. J Toxicol Environ Health Sci 3(3):65–73

Mamtaz R, Bache DH (2001) Reduction of arsenic in groundwater by coprecipitation with iron. Aqua—J Water Supply 50(5):313–322

Masrur AA et. al (2011) Evaluation of socio-economic impact of arseniccontamination in Bangladesh. J Toxicol Environ Health Sci 3(10):298–307. Available online at http://www.academicjournals.org/JTEHS. ISSN 2006-9820 ©2011 Academic Journals

Mitra et al (2002) Arsenic-related health problems among hospital patients in Southern Bangladesh. J Health Popul Nutr 20(3):198–204

Milton AH (2003) Health effects of arsenic- toxicity, clinical manifestation and health management in arsenic contamination: Bangladesh perspective. In: Feroze Ahmed M (ed) ITN Bangladesh. Centre for Water Supply and Waste Management, BUET, Dhaka, Bangladesh, pp 281–297

Moon KA et al (2013) Association between exposure to low to moderate arsenic levels and incident cardiovascular disease. A prospective cohort study. Ann Intern Med 159(10):649–659

Mostafa MG, Cherry N (2013) Arsenic in drinking water and renal cancers in rural, Bangladesh. Occup Environ Med 70(11):768–773

Mukherjee A et al (2006) Groundwater arsenic contamination: a global perspective with special emphasis to Asian scenario. J Health Popul Nutr 24(2):142–163

Nahar N, Hossain F, Hossain MD (2008) Health and socioeconomic effects of groundwater arsenic contamination in rural Bangladesh: new evidence from field surveys. J Environ Health 70 (9):42–47

Navas-Acien A et al (2008) Arsenic exposure and prevalence of type 2 diabetes in US adults. JAMA 300(7):814–822

NCIWRD, Report of the Working Group on Water Resources for the XIth Five Year Plan, KPMG Analysis (2010) https://www.kpmg.de/docs/Water_sector_in_India.pdf. Accessed 16 Sept 2014. Online Data base: http://en.wikipedia.org/wiki/Beedi. Accessed on 10 Oct 2011 at 22:36 Hrs. http://en.wikipedia.org/wiki/Rajmahal_Hills. Accessed at 28 Nov 12 at 20:45 Hrs. http://en.wikipedia.org/wiki/Gorubathan. Accessed at 28 Nov 12 at 21:02 Hrs. http://www.narc.org.np/rice_knowledge_bank/factsheet/boro.pdf. Accessed on 02 Aug 2015 at 14:03 Hrs. http://www.wbagrimarketingboard.gov.in/index.html. Accessed on 11 Nov 12 at 20:00 Hrs

Public Health Engineering Directorate (Government of West Bengal): Supplementary notes in connection with budget speech of Minister-in-Charge, PHED for the year: 2001, 2002, 2003, 2004, 2005, 2006, 2007

Rahman MM, Sengupta MK, Chowdhury UK et al D (2006) Arsenic contamination incidents around the World. In: Ravi N et. al (ed) Managing arsenic in the environment: from soil to human health. CSIRO Publishing, Australia, pp 3–30

Reports of School of Environmental Studies (SOES), Jadavpur University (May, 1991) Groundwater Arsenic Contamination Episode in Five Districts of West Bengal (A Priliminary study)

Reports of School of Environmental Studies (SOES), Jadavpur University (March 1994) Arsenic in Groundwater in Six Districts of West Bengal: The Biggest Calamity in the World: Priliminry Report of South 24-Parganas (Till Feb 1994)

Reports of School of Environmental Studies (SOES), Jadavpur University (September 1999) Field testing kits for arsenic—how effective are the million-dollar projects?

Reports of School of Environmental Studies (SOES), Jadavpur University (June 2000) Performance of arsenic removal plants for treating groundwater with community participation

Reports of School of Environmental Studies (SOES), Jadavpur University (September 2001) Fate of three crore-rupee arsenic removal plants in Murshidabad, West Bengal-India

Reports of School of Environmental Studies (SOES), Jadavpur University (January 2003) A study report on efficiency of 259 arsenic treatment plants installed in ten blocks of the districts, North 24-Parganas, West Bengal

Reports of School of Environmental Studies (SOES), Jadavpur University (March, 2003) Follow-up study of arsenic treatement plants in Murshidabad, West Bengal, India

Reports of School of Environmental Studies (SOES), Jadavpur University (December 2003) Usefulness of arsenic removal plants: a case study in Kolsur Gram Panchayat of Deganga Block in North 24-Parganas, West Bengal, India (fifth report)

Reports of School of Environmental Studies (SOES), Jadavpur University (July 2004) 6th report on arsenic removal plants (ARPs) installed in the arsenic affected villages of West Bengal

Reports of and School of Environmental Studies (SoES), Jadavpur University and Dhaka Community Hospital (DCH), Bangladesh (February 2010) Are millions in Ganga-Meghna-Brahmaputra (GMB) Plain already exposed to arsenic contaminated water potentially at risk from cancer? A preliminary follow up study on arsenic patients registered during 1995 to 2000

Reports of Public Health Engineering Directorate (PHED), Government of West Bengal (1994) Report on arsenic pollution in groundwater in West Bengal

Reports of Public Health Engineering Directorate (PHED), Government of West Bengal (1995) Perspective plan for arsenic affected districts of West Bengal

Reports of Public Health Engineering Directorate (PHED), Government of West Bengal (2006) Master plan to tackle arsenic contamination in West Bengal

Reports of Public Health Engineering Directorate (PHED), Government of West Bengal (2006, 2007, 2008) Arsenic contamination in groundwater of West Bengal—strategy and action taken

Reports of Public Health Engineering Directorate (PHED), Government of West Bengal (2007) Status of rural piped water supply schemes

Reports of Public Health Engineering Directorate (PHED), Murshidabad Division, Government of West Bengal (2003, 2007, 2008) Activities of PHED in Murshidabad District

Roy J (2008) Economic benefits of arsenic removal from groundwater—a case study from West Bengal, India. Sci Total Environ 397:1–3

Roy J, Das A (2009) Revisiting actions for arsenic safe drinking water supply in rural West Bengal. In: Proceeding of the regional workshop on groundwater related issues—West Bengal, Andaman & Nicobar Islands and Sikkim at NTHA, Salt Lake, Kolkata, 13th & 14th Feb 2009 organised by CGWB (Eastern Region), Ministry of Water Resources, GoI

Roy J, Das A, Dasgupta S (2015) Arsenic safe water access system (ASWAS): missing links. In: Proceedings of the workshop on the Incidence of High Arsenic in Groundwater in West Bengal and its remedial measures, Government of India, Ministry of Water Resources, River Development and Ganga Rejuvenation, Kolkata, 24 Mar 2015

Saha KC, Chakraborti D (2001) Seventeen years experience of arsenicosis in West Bengal, India. In: Chappell et al (ed) Arsenic exposure and health effects, Elsevier, Amsterdam, pp 387–396

Sanyal SK, Nasar SKT (2002) Arsenic contamination of groundwater in West Bengal (India): build-up in soil-crop systems. In: Proceeding of the international conference on water related disasters, Kolkata, 5–6 Dec 2002

Sarker MMR (2010) Determinants of arsenicosis patients' perception and social implications of arsenic poisoning through groundwater in Bangladesh. Int J Environ Res Public Health 7:3644–3656

SEU (2004) Social exclusion unit (Office of the Deputy Prime Minister), breaking the cycle: taking stock of progress and priorities for the future (London). http://webarchive.nationalarchives.gov.uk/+/http:/www.cabinetoffice.gov.uk/media/cabinetoffice/social_exclusion_task_force/assets/publications_1997_to_2006/breaking_report.pdf. Accessed 14 July 2014

Sengupta M K (2006) Situation of groundwater arsenic contamination problem and human sufferings in West Bengal (Unpublished doctoral dissertation). Jadavpur University, Kolkata, West Bengal (India)

Shah T, Makin I, Sakthivadivel R (2005) Limits to leapfrogging: issues in transposing successful river basin management institutions in the developing world. In: Svendsen M (ed) Irrigation and river basin management: options for governance and institutions. CAB international, Wallingford, pp 31–49

Shah T (2007) The groundwater economy of South Asia: an assessment of size, significance and socio-ecological impacts. In: Giordano M, Villholth KG (eds) The agricultural groundwater revolution: opportunities and threats to development. CAB International, pp 7–36. http://ageconsearch.umn.edu/bitstream/158009/2/H039669.pdf. Accessed 10 May 2014

Sharma AK (2006) Arsenic removal from water using naturally occurring Iron and the associated benefits on health in affected regions. Ph. D. thesis, Institute of Environment & Resources, Technical University of Denmark (Unpublished).

Sorg TJ, Logsdon GS (1978) Treatment technology to meet the interim primary drinking water regulations for inorganics: part 2. J Am Water Works Assoc 70(7):379–393

Smith AH, Hira Smith MM (2004) Arsenic drinking water regulations in developing countries with extensive exposure. Toxicology 196:39–44

Sultana S, Hossain QZ, Pervin R (2012) Socioeconomic condition and health status of chronic arsenicosis patients in Jessore, Bangladesh. Int J Adv Nutr Health Sci 1(1):9–17

Syed EH, Poudel KC et al (2012) Quality of life and mental health status of arsenic-affected patients in a Bangladeshi population. J Health Popul Nutr 30(3):262–269

Tseng W-P, Chu H-M, How S-W et al (1968) Prevalence of skin cancer in an endemic area of chronic arsenicism. Taiwan J Nat Cancer Inst 40:453–462. http://www.ncbi.nlm.nih.gov/pubmed/5644201. Accessed 16 June 2014

UNDP—Human Development Report (2006) Beyond scarcity: power, poverty and the global water crisis, Published for the United Nations Development Programme

Vather M (1994) Differences in the metabolism of arsenic compounds. Appl Organomet Chem 8:175–182

WaterAid (2008) Safe drinking water in rural areas community based approaches .In: Conference background paper and abstracts, New Delhi, 8–10 April 2008

WHO (World Health Organization) (1958) International standard for drinking water

WHO (World Health Organization) (1963) International standard for drinking water

WHO (World Health Organization) (1971) International standard for drinking water

WHO (World Health Organization) (1993) Our planet, our health: report of the WHO Commission on Health and Environment, World Health Organization, Geneva, pp 458–467 (Chapter on Water)

WHO (World Health Organization) (1999) The world health report. WHO, Geneva

WHO (World Health Organization) (2000) Sustainable development and healthy environments. Towards an assessment of the socioeconomic impact of arsenic poisoning in Bangladesh

WHO (World Health Organization) (2004) Guidelines for drinking water quality, 3rd ed, vol 1. Recommendation, Geneva

World Health Organisation (1994): Financial of water supply and sanitation: a handbook, Geneva. ISBN 9241544724, 9789241544726, 83p

WSSCC, Resource Pack on the Water and Sanitation Millennium Development Goals (2004)

Additinal Reading List

Ahmed MF (2003) Laboratory methods for the analysis of arsenic in drinking water in arsenic contamination: Bangladesh perspective. In: Feroze Ahmed M (ed) ITN Bangladesh, Centre for Water Supply and Waste Management, BUET, Dhaka, Bangladesh, pp 481–509

Ahmad J, Goldar BN, Misra S, Misra, S et al (2003) Fighting arsenic, listening to rural communities: willingness to pay for arsenic-free, safe drinking water in Bangladesh. Water and Sanitaton Programme. www.wsp.org/publications/sa_arsenic_learning.pdf

Ahmed S, Mukherjee A, Nayek B et al (2007) Groundwater Arsenic Contamination and it health effects in Ganga-Meghna-Brahmaputra (GMB) plain and its surrounding areas: a nineteen years study. Chin J Epidemiol 26(1):43–47

Alberini Anna, Kruprick Alan (2000) Cost-of-illness and willingness to pay estimates of the benefits of improved water quality: evidence from Taiwan. Land Econ 76(1):37–53

Amrose ES, Bandaru SRS, Dlaire C et al (2014) Electro-chemical arsenic remediation: field trials in West Bengal. Sci of Total Environ 488–489:539–546

Aposhian H.V (2000) DMP Increases the urinary excretion of arsenic in human. In: Proceedings of 4th international conference on arsenic exposure and health effects, San Diego, 18–22 June 2000

Timothy Bartik J (1988) Evaluating the benefits of non-marginal reductions in pollution using information on defensive expenditure. J Environ Econ Manage 15:11–127

Bhattacharya P, Chatterjee D, Jacks G (1997) Occurrence of arsenic contaminated groundwater in alluvial aquifer from delta plains, Eastern India: option for safe water supply. Water Resour Develop 13:9–92

Chakraborti D, Basu GK, Biswas BK et al (2001) Characterization of arsenic-bearing sediments in the Gangetic delta of est Bengal, India. In: Chappell WR, Abernathy CO, Calderon RL (eds) Arsenic exposure and health effects. Elsevier, Amsterdam, pp 27–52

Chakraborti D, Rahaman MM, Paul K et al (2002) Arsenic calamity in India and Bangladesh sub-continent—whom to blame? Talanta 58:3–22

Chakraborty D, Hussam A, Alauddin M (2003) Foreword. J Environ Sci Health 38(1):xi–xv

Chakraborti D, Sengupta M, Rahman M.M, Chowdhuri et al (2003) Groundwater arsenic exposure in India. In Chappell et.al (ed), Arsenic exposure and health effects, Elsevier, Amsterdam-Lausanne-New York-Oxford-Tokyo, p 3-24

Chakraborti D, Mukherjee SC, Saha KC et al (2003) Arsenic toxicity from homeopathic treatment. J Toxicol 41(7):963–967

Chakraborty D, Sengupta MK, Rahaman MM et al (2004) Groundwater arsenic contamination and its health effects in the Ganga-Meghna-Brahmaputra plain. J Environ Monitor 6:75N–83N

Chakraborti I, Sinha RK, Ghosh K (2003) Arsenic in food chain—a case study on both raw and cooked food in arsenic contamination: Bangladesh perspective. In: Feroze Ahmed M (ed) ITN Bangladesh, Centre for Water Supply and Waste Management, BUET, Dhaka, Bangladesh, pp 227–240

Chakraborti D, Rahaman MM, Mitra S et al (2013) Groundwater arsenic contamination in India: a review of its magnitude, health, socio-economic effects and approaches for arsenic mitigation. J Ind Soc Agric Stat 67(2):1–9

Cheng CR, Liang S, Wang HC et al (1994) Enhanced coagulation for arsenic removal. J Am Water Works Assoc 86(9):79–90

Chowdhury TR, Basu GK, Mandal BK et al (1999) Arsenic poisoning in the Ganges delta. Nature 401:545–546

Chowdhury UK, Biswas BK, Roychowdhury T et al (2000) Groundwater arsenic contamination in Bangladesh and West Bengal, India. Environ Health Perspect 108(5):393–397

Chowdhury UK, Rahaman MM, Mondal BK et al (2001) Groundwater arsenic contamination and human suffering in West Bengal—India and Bangladesh. Environ Sci 8(5):393–415

Choudhury MAI, Uddin MT, Ahmed MF et al (2006) Collapse of socio-economic base of Bangladesh by arsenic contamination in groundwater. Pak J Biol Sci 9(9):1617–1627

Cournat PN, Porter RC (1981) Averting expenditure and the cost of pollution. J Environ Econ Manage 8:321–329

Das A (2008) Health hazards caused by groundwater arsenic pollution in rural West Bengal, Kurukshetra. J Rural Develop 56(12):34–37

Das D, Chatterjee A, Samanta G, Chowdhury TR et al (2001) A simple household device to remove arsenic from groundwater and two years performance report of arsenic removal plant for treating groundwater with community participation. In: Feroze Ahmed M, Ashrafi Ali M, Zafar Adeel M (eds) Proceedings of BUET-UNU international workshop on technologies for arsenic removal from drinking water, Dhaka, Bangladesh

Das D, Samanta G, Mandal BK, Chowdhury TR et al (1996) Arsenic in groundwater in six districts of West Bengal, India: biggest calamity in the world. Environ Develop Health 18:5–15

Dhar RK, Biswas BK, Samanta G et al (1997) Groundwater arsenic calamity in Bangladesh. Curr Sci 73:48–59

Dutta DV, Kaul MK (1976) Arsenic content of tubewell water in villages in northern India: a concept of arsenicosis. J Assoc Physicians India 24:599–604

EPA (U.S. Environmental Protection Agency) (1988) Special report on ingested inorganic arsenic: skin cancer, Nutritional Essentiality, EPA/625/3-87/013. Washington, DC

Erickson B (2003) Field kits fail to provide accurate measure of arsenic in groundwater. Environ Sci Technol 37(1):35A–38A

Ashok Gadgil (1998) Drinking water in developing countries. Annual review. Energy Environ 23:253–286

Goldar B, Mishra S (2001) Valuation of environmental goods: correcting for bias in contingent valuation studies based on willingness to accept. Am J Agric Econ 83(1):150–156

Guha Mazumdar DN (2001) Clinical aspects of chronic arsenic toxicity. J Assoc Phys India 49:650–655

Hanchett S (2004) Social aspects of the arsenic contamination of drinking water: a review of knowledge and practice in Bangladesh and West Bengal. Report for the Arsenic Policy Support Unit, Local Government Division, Government of Bangladesh

Hanley N, Shogren JF, White B (1997) Environmental economics in theory and practice. Oxford University Press, Oxford

Harvey CF, Swartz CH, Badruzzaman ABM et al (2002) Arsenic mobility and groundwater extraction in Bangladesh. Science 98:1602–1606

Hassan MM, Atkins PJ, Dunn CE (2005) Social implications of arsenic poisoning in Bangladesh. Soc Sci Med 61(10):2201–2211

Harrington Winston, Portney Paul R (1987) Valuing the benefits of health and safety regulations. J Urban Econ 22(1):101–112

Hering JG, Chen PY, Wilkie JA et al (1996) Arsenic removal by ferric chloride. J. Am Water Works Assoc 88(4):155–167

Hironka H, Ahmed SA (2003) Arsenic concentration of rice in Bangladesh, in fate of arsenic in the environment In: Proceedings of the BUET-UNU international symposium, Dhaka, 5–6 Feb 2003

Amir Hossain M, Sengupta MK, Ahamed Sad et al (2005) Ineffectiveness and poor reliability of arsenic removal plants in West Bengal, India. Environ Sci Technol 39:4300–4306

Hossain M Amir, Mukherjee A, Sengupta MK et al (2006) Million dollar arsenic removal plants in West Bengal, India: useful or not? Water Qual Res J Can 41(2):216–225

Hsueh YM, Cheng GS (1995) Multiple risk factors associated with arsenic-induced skin cancers, effects of chronic liver diseases and malnutritional status. British Journal of Cancer 71:109–114

Hussain A, Iftikhar ZM (2003) Arsenic hazards and possible public health management in arsenic contamination: Bangladesh perspective. In Feroze Ahmed M (ed) ITN Bangladesh. Centre for Water Supply and Waste Management, BUET, Dhaka, Bangladesh, pp 268–280

ICRISAT, Annual Rainfall, http://www.icrisat.org/text/research/nrmp/dfid/text. Accessed on Jan 2007

Islam FS, Gault AG, Boothman C et al (2004) Role of metal reducing bacteria in arsenic release from Bengal delta sediments. Nature 430:68–71

Joshi A, Chaudhury M (1996) Removal of arsenic from groundwater by iron oxide-coated sand. ASCE J Environ Eng 122(8):769–771

Kadekodi Gopal K (2004) Environmental economics in practice. Oxford University Press, New Delhi

Kartinen EO, Martin CJ (1995) An overview of arsenic removal processes. Desalination 103(1–2):79–88

Khan MZH (2007) Managing the arsenic disaster in water supply: risk measurement, costs of illness and policy choices for Bangladesh. South Asian Network for Development and Environmental Economics (SANDEE) Working Paper No. 27-07. Available on website www.sandeeonline.org

Charles Kolstad D (2000) Environmental economics. Oxford University Press, New York

Kondo H, Ishiguro Y, Ohno K et al (1999) Naturally occurring arsenic in the groundwater in the Southerrn region of Fukunoka Prefecture, Japan. Water Resour 33:1967–1972

Lewis DR, Southwick JW et al (1999) Drinking water arsenic in Utah: a cohort mortality study. Environ Health Perspect 107:359–365

Mandal BK, Roy Choudhury T, Samanta G et al (1997) Chronic arsenic toxicity in West Bengal. Curr Sci 72(2):114–117

Mandal BK, Chowdhury TR, Samanta G et al (1998) Impact of safe water for drinking and cooking on five arsenic-affected families for 2 years in West Bengal, India. Sci Total Environ 218:185–201

Jakariya M, Chowdhury AMR, Hossain Z et al (2003) Sustainable community based safe water options to mitigate the Bangladesh arsenic catastrophe—an experience from two upazilas. Curr Sci 85(2):141–146

Meharg AA (2004) Arsenic in rice-understanding a new disaster for South East Asia. Trends Plant Sci 9:415–417

Mitra SR, Guha Majumdar DN, Basu A et al (2004) Nutritional factors and susceptibility to arsenic—caused skin lesions in West Bengal. Environ Health Perspect 112(10):1104–1109

Mukherjee A, Sengupta M, Hossain MA (2006) Groundwater arsenic contamination: a global perspective with special emphasis to Asian scenario. J Health Popul Nutr 24(2):142–163

Mukherjee A, Sengupta MK, Hossain MA et al (2005) Are some animals more equal than others? Toxicology 208:165–169

Mukherjee A, Sengupta MK, Ahamed S et al (2005) Comment on reliability of a commercial kit to test groundwater for arsenic in Bangladesh. Environ Sci Technol 39(14):5501–5502

Nasreen M (2002) Social impact of arsenicosis of arsenic contamination in Bangladesh. In: Ahmed MF (ed) ITN Bangladesh, pp 340–353

Nicolli HB, Suriano JM, Gomez Peral MA et al (1989) Groundwater contamination with arsenic and other trace elements in an area of the Padma Province of Corboda, Argentina. Environ Geol Water Sci 14:3–16

Norra S, Aggrawala P, Wagner F et al (2005) Impact of irrigation with arsenic rich groundwater on soil and crops: a geochemical case study in Malda district, West Bengal. App Geochem 20:1890–1906

Oshikawa S, Geater A, Chongsuvivat Wong V et al (2001) Long term hanges in severity of arsenical skin lesions following intervention to reduce arsenic exposure. Environ Sci 8:435–448

Ostro Bart D (1983) The effects of air pollution on work loss and morbidity. J Environ Econ Manage 10:371–382

Pal A, Chowdhury UK, Modal D et al (2009) Arsenic burden from cooked rice in the populations of arsenic affected and non affected areas and Kolkata city in West-Bengal, India. Environ Sci Technoly 43(9):3335–3335

Pal A, Chowdhury UK, Modal D et al (1996) Calcutta's air quality during winter with special reference to heavy metals and development of analytical techniques for lead, mercurry and arsenic by Gautam Samanta Rahman submitted at Jadavpur University

Pal A, Chowdhury UK, Modal D et al (2003) Present status of groundwater arsenic contamination in Bangladesh and detailed study of Murshidabad, one of the affected neighboring districts in West Bengal-India by Mohammad Mahmudur Rahman submitted at Jadavpur University

Pal A, Chowdhury UK, Modal D et al (2006) Situation of groundwater arsenic contamination problem and human sufferings in West Bengal by Mrinal K Sengupta submitted at Jadavpur University

Pal A, Chowdhury UK, Modal D et al (2007) Sustainability, water use and economic incentives—a case for greater Kolkata by Monojit Ghosh submitted at Jadavpur University

Pal A, Chowdhury UK, Modal D et al (2010) Investigation for safe water options in groundwater arsenic contaminated rural and urban settlement in lower Ganga Plain, West Bengal, India by Bhaskar Das at Jadavpur University

Public Health Engineering Directorate (Government of West Bengal): Budget speech of honorable minister-in-charge in respect of demand no. 45 for the years 2005-06, 2006-07

Quamruzzaman Q, Mahmuder R, Khondeker A (2003) Effects of arsenic on health in arsenic contamination: Bangladesh perspective. In Feroze Ahmed M (ed) ITN Bangladesh. Centre for Water Supply and Waste Management, BUET, Dhaka, Bangladesh, pp 257–267

Rahman MM, Chowdhury UK, Mukherjee SC et al (2001) Chronic arsenic toxicity in Bangladesh and West Bengal-India—a review and commentry. J Toxicol Clin Toxicol 39(7):683–700

Rahman MM, Mukherjee D, Sengupta MK et al (2002) Effectiveness and reliability of arsenic field testing kits: are the million dollar screening projects effective or not? Environ Sci Technol 36 (24):5385–5394

Rahman MM, Sengupta MK et al (2005) Murshidabad - one of the nine groundwater arsenic affected districts of West Bengal, India, Part I and Part II. J Toxicol Clin Toxicol 43:823–834

Rahman MM, Sengupta MK et al (2005) The magnitude of arsenic contamination in groundwater and its health effects to the inhabitants of the Jalangi- one of the 85 arsenic affected blocks in West Bengal. Sci Total Environ 338(3):189–200

Rahaman MM, Sengupta MK et al (2005) Status of groundwater arsenic contamination and human suffering in a Gram Panchayet (cluster of villages) in Murshidabad, one of the nine arsenic affected districts in West Bengal, India. J Water Health 3(3):283–295

Ratna Reddy V, Kullappa M, Mohan Rao D (2008) Social cost-benefit analysis of improved water quality in rural areas: an exploratory study in coastal AP. J Soc Econ Develop 10(1):69–97

Ravenscroft P, Brammer H, Richards K (2009) Arsenic pollution—a global synthesis. RGS-IBG Book Series, Wiley-Blackwell, New York

Rawal V, Swaminathan M (1998) Changing trajectories- agricultural growth in West Bengal 1950-1996. Econ Polit Weekly 33(40):2595–2602

Report of National Institute of Hydrology (NIH), Roorkee Central Ground Water Board (CGWB), New Delhi Under the aegis of: Ministry of Water Resources Government of India New Delhi, (June 2010) Mitigation and remedy of groundwater arsenic menace in India: a vision document". Available at http://www.nih.ernet.in/Docs/mitigationandremedy.pdf. Accessed on 05 Sept 2015

Reports of School of Fundamental Research (SFR) (2003) Lessons from arsenic technology park and critical issues of arsenic mitigations

ROHC (2006) Clinico-epidomiological study of arsenic exposed population in West Bengal. Regional Occupational Health Center (Eastern), Indian Council of Medical Research, Kolkata

Roy Chowdhury T, Basu GK, Mandal BK et al (1999) Arsenic poisoning in the Ganges delta. Nature 401:545–546

Roy J, Chattopadhyay S, Mukherjee S et al (2004) An economic analysis of demand for water quality, Case of Kolkata. Econ Polit Weekly 186–192.

Roy J, Sohini S (2004) Willingness to pay studies: a policy tool. In: Abstracts of the conference on market development of water and waste technologies through environmental economics, May, Paris. http://www.cerna.ensmp.fr

Roy J, Chottapadhyay S, Mukerjee S et al (2004) An economic analysis of demand for water quality: a case from Kolkata City. Econ Polit Weekly XXXIX(2):186–192

Saha K C (2000) Malignancy in Arsenicosis. In: Proceedings of the international workshop on control of arsenic contamination in groundwater, Public Health Engineering Department, Government of West Bengal, Kolkata, India, 5–6 Jan 2000

Saha K C (2003) Saha's grading of arsenicosis progression and treatment. In: Chappell WR, Abernathy CO, Calderon RL,Thomas DJ (eds) Arsenic exposure and health effects. Elsevier, Amsterdam, pp 391–414

Saha KC (1999) Arsenic in tubewell: disease and remedy. Sadananda Prashani, Kolkata

Saha KC (2004) Arsenicosis in West Bengal (environmental problems and solutions). Sadananda Prakshani, Kolkata

Sankar U (2001) Environmental economics. Oxford University Press, New Delhi

Sanok WJ, Ebel JG, Manzell KL et al (1995) Residues of arsenic and lead in potato soils on Long Island. Chemosphere 30(4):803–806

Sarkar A, Mehrotra R (2005) Social dimensions of chronic arsenicosis in West Bengal (India). Epidemiology 16(5):68

Sarkar A (2009) Sustainable solutions to arsenic contamination of groundwater: The Ganga-Meghna-Brahmaputra Basin. In Pascual U et.al (eds) Water, agriculture, and sustainable well-being. Oxford University Press, Oxford, pp 72–92

Shen YS (1973) Study of arsenic removal from drinking water. J Am Water Works Assoc 65 (8):543–548

Shrestha RR, Surestha MP, Upadhyay, NP et al (2003) Groundwater arsenic contamination, its health impact and mitigation program in Nepal. J Env Sci Health Part A Toxic/Hazard Subst Env Eng A 38(1):185–200

Smith AH, Lingas EO, Rahaman M (2000) Contamination of drinking water of arsenic in Bangladesh. A public healthy emergency. Bull World Health Organ 78(9):1093–1103

Sun D, Liy Y, Pi J, Sun Y et al (2006) current research problems of chronic arsenicosis in China. J Health Popul Nutr 24(2):176–181

The World Bank (1999) Entering the 21st century: "world development report", (1999-2000). Oxford University Press, New York

The World Bank (1995) Toward sustainable management of water resources. The World Bank, Washington

The World Bank (1992) The World Development Report. Oxford University Press, New York

Tandulkar N, Bhattacharya P, Mukherjee A B (2001) Preliminary assessment of arsenic contamination in groundwater in Nepal. In: Proceedings of international conference on arsenic in the Asia-Pacific region: managing arsenic for our future, Book of Abstracts, CSIRO Land & Water, Adelaide, South Australia, pp 103–105

Takamatsu T, Aoki H, Yoshida T (1982) Determination of arsenate, arsenite, monomethyl arsonate and dimethylarsinate in soil polluted with arsenic. Soil Sci 133(4):239–246

Tam GKII, Charboneau SM, Bryce F, Pomroy C, Sandi E (1994) Metabolism of inorganic arsenic (^{74}As) in humans following oral ingestion. Toxicol Appl Pharmacol 50:319–22

Von Ehrenstein OS, Guha Mazumder DN, Smith MH et al (2006) Pregnancy outcomes, infant mortality, and arsenic in drinking water in West Bengal, India. Am J Epidemiol 163:662–669

Welch AH, Lico MS, Hughes JL (1988) Arsenic in groundwater of the western United States. Ground Water 26(3):333–347

Willams PN, Pricea AH, Raab A et al (2005) Variation in arsenic speciation and concentration in paddy rice related to dietary exposure. Environ Sci Technol 39:5531–5540

Websites

www.wbphed.gov.in
www.soesju.org
www.cgwber.nic.in
www.whoindia.org
www.dngmresfoundation.org
www.planningcommission.org
www.watersupply.com
www.waterresourcesdevelopment.com
www.suatainability.com
www.safedrinkingwater.com
www.wtp.com
www.wateraidindia.org
www.ddws.nic.in
www.censusindia.net
www.projectwellusa.org

Index

A. Das et al., *Socio-Economic Analysis of Arsenic Contamination of Groundwater in West Bengal*, India Studies in Business and Economics,
DOI 10.1007/978-981-10-0682-1

The manufacturer's authorised representative in the EU is Springer Nature Customer Service Centre GmbH, Europaplatz 3, 69115 Heidelberg, Germany. If you have any concerns regarding our products, please contact ProductSafety@springernature.com

Printed and bound by CPI Group (UK) Ltd, Croydon, CR0 4YY
30/09/2026
02225467-0017